D1214710

Lighting
the
Workplace

Lighting the Workplace

BY THE EDITORS OF PBC INTERNATIONAL

PBC International, Inc. ■ New York

749.63
L626P

PACKARD LIBRARY

FEB 21 1989

OF THE COLUMBUS COLLEGE OF ART AND DESIGN

Distributor to the book trade in the
United States and Canada:

Rizzoli International Publications, Inc.
597 Fifth Avenue
New York, NY 10017

Distributor to the art trade in the United
States:

Letraset USA
40 Eisenhower Drive
Paramus, NJ 07653

Distributor to the art trade in Canada:

Letraset Canada Limited
555 Alden Road
Markham, Ontario L3R 3L5, Canada

Distributed throughout the rest of the
world by:

Hearst Books International
105 Madison Avenue
New York, NY 10016

Copyright © 1988 by PBC
International, Inc.
All rights reserved. No part of this book
may be reproduced in any form
whatsoever without written permission
of the publisher. For information about
this book write: PBC International, One
School Street, Glen Cove, NY 11542.

Library of Congress Cataloging-in-Publication Data

Lighting the workplace.

Bibliography: p.
Includes index.
1. Office buildings—Lighting.   I. Title.
TR4399.035J26   1988   621.32'2523   87-43302
ISBN 0-86636-062-X

Color separations, printing and binding by
Toppan Printing Co. (H.K.) Ltd.

Typesetting by
Jeanne Weinberg Typography

PRINTED IN HONG KONG
10 9 8 7 6 5 4 3 2 1

MANAGING DIRECTOR	Penny Sibal-Samonte
CREATIVE DIRECTOR	Richard Liu
FINANCIAL DIRECTOR	Pamela McCormick
ASSOCIATE ART DIRECTOR	Daniel Kouw
EDITORIAL MANAGER	Kevin Clark
ARTISTS	William Mack
	Kim McCormick

CONTENTS

INTRODUCTION 8

by Gary R. Steffy, IALD, IES

INTRODUCTION

Gary R. Steffy,

IALD, IES

What is "good lighting" for the workplace? The best lighting for the workplace isn't even for the workplace. It's for the *people* in the workplace. Lighting for people. The best architecture and the best office equipment are pretty sad indeed if the lighting for the users of that architecture and equipment isn't appropriate to their needs and wants.

Good lighting is a synthesis of a host of factors, including visual tasks, visual comfort, architectural integration, environmental systems integration and psychological considerations. There is no magic to good lighting. There is no panacea. You will see in this collection a variety of installations which meet a number of criteria. No single luminaire, lamp, surface color or finish provided the designer with the "right" solution. A team effort among owner, user, architect, interior designer, electrical and mechanical engineers and manufacturers produces a "right" solution for their set of circumstances. Much like the caveat in television ads for racy autos: "Expert designer at the board. Do not attempt this without appropriate expertise, hard work and team effort."

You'll see that esoteric lamps (a.k.a. bulbs) are not required, nor are custom luminaires always the answer. While opulent budgets may make a project "easier," you'll see that some of the best solutions use off-the-shelf equipment in clever ways. At the same time, you should be prepared to put the lighting budget in perspective.

Budgets. Many of us shudder at the term. Still others allot next to nothing in the hopes that the designer will make something of nothing. Many argue that lighting is just like any other building system. But it's much different. Even the most opulent lighting is usually less than 10 percent of the total building budget. Most well-designed, fiscally responsible lighting is approximately 5 percent of the total building budget (of course, this depends on the size of the project—larger projects generally have a smaller percentage of the total budget allocated to lighting, and small projects have a greater percentage).

Check other systems' budgets. What percentage of the project goes to other systems? Yet lighting is responsible for nearly all of the information we receive and our subsequent response to that information in the workplace. How well we "receive and respond" depends upon how well the lighting system is designed.

Think of it another way. The lighting will probably "live" untouched for 15 years in a building. Over that same 15-year period think about how much money will go toward salaries and benefits. Now, think of the lighting system as an investment toward maximizing the output derived from those salaries and benefits. Lighting *is* an investment in both people and architecture.

As the high-tech age comes into its own, we find that we have been designing more for the machines than the people. While the machines may provide the *potential* for productivity gains, *people* make it happen. The high-tech office gadgetry combined with the energy crunches to make us aware of our lack of sensitivity to the people in the environment.

This sensitivity to people will help to continue the increased awareness and importance of lighting and lighting design. The work herein exemplifies this growing awareness. The workplace is no longer an environment that shows off rows and rows of machines. Illuminances (quantities of light falling on surfaces) are no longer in vogue. Luminances (measured brightnesses—the light we actually see), luminance ratios (contrasts) and chromatic (color) contrast

are the recognized quantitative factors. Subjective impressions (*e.g.*, pleasantness, spaciousness, relaxation) are the recognized qualitative factors. Just as the baby boomers have impacted cars (remember the bare little cars of ten years ago) and foods (remember when meat and potatoes were ''it''), they will impact the workplace. For the designer, this means attention to the *people* in the space.

Lamp and luminaire technology, while seemingly off track during the energy crisis, has provided us with some very exciting equipment. Lamps continue to move toward a more compact size, with improvement in efficiency, source color and color rendering. Just like cars, lamps have become more personable, flattering and efficient. These compact lamps have led to smaller, more human scale luminaires. Indirect lighting is now efficient *and* visually unobtrusive. Direct lighting can now be more delicate, better controlled *and* efficient.

Lamps will continue to shrink. Look for smaller sizes, lower wattages, and continued color and efficiency improvements. Over the next decade we are likely to see an incandescent-like lamp with life and efficiencies nearing that of fluorescent lamps. Luminaires will be introduced which are much more appropriate to human-scale than the 2 ft. x 4 ft. troffer could ever be. Non-linear indirect lighting without the harshness or starkness of traditional fluorescent or metal halide lamps will be introduced. Soft, visually pleasant lighting systems will put the human touch back into the high-tech workplace.

Some tasks may change over the next decade. Voice recognition technology may arrive in the workplace—speaking to and listening to a computer may be commonplace. Investigations into and experiments with various CRT screens may lead to the development of screens which require ambient light for visibility of the screen. Of course, the computers will no doubt continue to spew ream after ream of paper, so paper tasks will still be common.

Lighting may change to respond to changing tasks, but lighting will still be the key to providing a comfortable, productive workplace. Please spend some time with the installations exhibited on the following pages. As Howard Brandston, lighting designer of the Statue of Liberty and many well-known corporate facilities, has so clearly pointed out to me, *observation* is the most important key to visualizing and designing environments for people. Use this book as your observation tower. But don't stop looking after the pages end.

Gary R. Steffy, IES, IALD, is president of Gary Steffy Lighting Design in Ann Arbor, MI. His firm, which has earned several lighting design awards, is a Sustaining Member of the Illuminating Engineering Society. Current clients include Steelcase Inc., Prudential Insurance Company, Stow & Davis, and the General Electric Company. Before establishing his own firm, Mr. Steffy held positions with the lighting group at Smith, Hinchman & Grylls, and with Owens-Corning Fiberglas. Mr. Steffy has taught at Penn State, Wayne State and Michigan State Universities. He guest

lectures at the University of Colorado. His work and articles have appeared in numerous magazines, including *Lighting Design + Application, Architectural Lighting, Architectural Record, Interior Design,* and *Building Design and Construction.* Mr. Steffy recently co-authored with John Flynn and Art Segil the second edition of ARCHITECTURAL INTERIOR SYSTEMS. He is currently president of the International Association of Lighting Designers.

CHAPTER 1
CORPORATE HEADQUARTERS

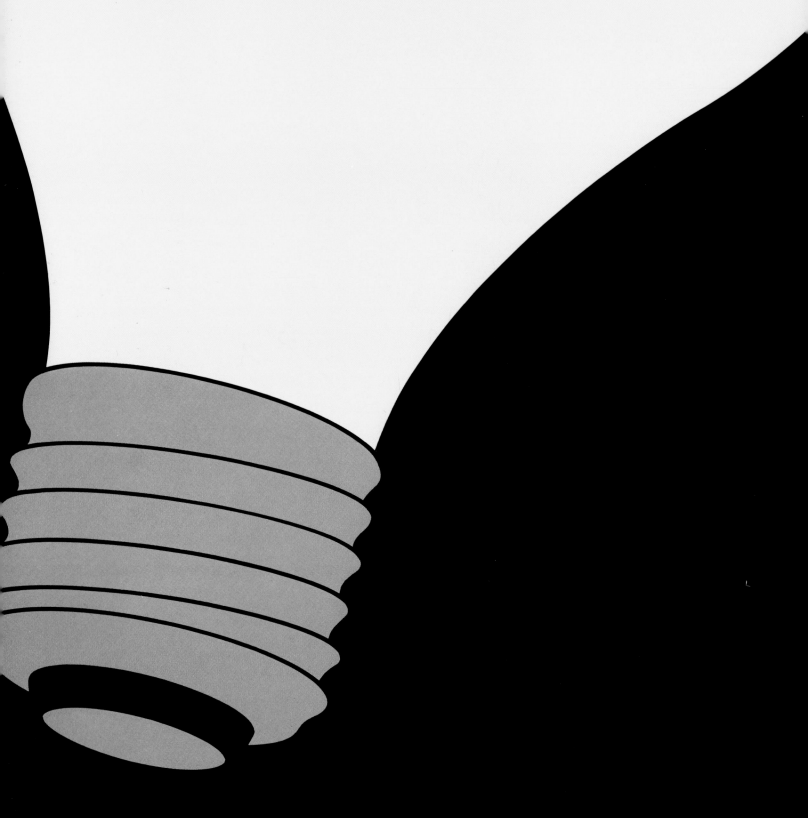

The projects selected for this chapter illustrate the importance of lighting to establish the impression a company portrays to the public.

Corporate headquarters house the executive offices, financial departments, upper management, and many other facilities. This building (or in some cases, buildings) serves as the "showcase" for a company.

This chapter features a variety of headquarters: from the beautiful Connecticut-based CIGNA Corporation (with its elegant atrium), to the sophisticated and sumptuous offices of National Westminster Bank located in New York City, to the recently remodeled U.S. News and World Report Building in Washington, D.C. Each of these projects had to address many design problems and concerns—ornamental lighting for lobbies, public areas and reception rooms; ambient lighting for hallways, offices, conference rooms, cafeterias and employee areas; task and function lighting for work areas, computer rooms and cubicles. All of these lighting problems were faced and resolved in these projects.

CIGNA Corporation

Project:	CIGNA Corporation
Location:	Bloomfield, CT
Lighting Consultants:	Raymond Grenald Associates
Architect:	The Architects Collaborative (TAC) Cambridge, MA
Interior Designer:	Interspace Incorporated Philadelphia, PA
Photographer:	Steve Rosenthal
Manufacturer of Lighting Equipment:	Peerless Electric Company

The new building on the rolling 65-acre campus was sited overlooking a spring-fed pond to the south of the original headquarters—distant enough to avoid a clash of styles, near enough to enhance the overall building ensemble—and oriented with major exposures to north and south for optimum sun control—the key to natural lighting, and so to energy efficiency. Held by code to four stories and by conviction to openness to daylight and views, the 534,000-square-foot structure was strung out in a footprint of 890 by 260 feet, with parallel offset wings on either side of a two-level atrium that TAC describes as the embodiment of its design philosophy.

Certainly it is the fundament of its design. In addition to amplifying natural light, the atrium serves as the social hub of the employee community and brings human scale to a building equivalent to a felled 80-story tower. It is also the circulation hub, linking the building floors via a central escalator and glass elevators at the ends of the court, and connecting its wings with flying bridges.

In the upper-level bays, central corridors alternate with balconies to add to the variety of visual and spatial experiences employees enjoy.

Though lacking the bravura of the atrium, the office floors are as skillfully planned, playing their crispness and polish against its varied textures, exuberant landscaping, and colorful adjunct spaces. Divided into four column-free pods arranged in mirror image around the atrium, the bright and airy working areas respond to the directive for unobstructed views via a remarkable custom-designed lighting/partition system. All interior walls parallel to the windows consist of post-free glass panels which float beneath a "ceiling" sketched in by light tubes that in fact support them.

The watch-like precision of detailing that marks the partition system is but the most striking example of the extra-ordinary attention to detail and the high level of materials and finish to be found throughout—and without—CIGNA's new headquarters.

Human scale, light and openness is architecturally expressed and provides a quality appearance appropriate to business and organizational functions.

The atrium's clear glass skylight (which can be shaded on the south as needed) pours daylight into the court and adjacent work areas, making feasible the 66-foot deep office bays that overlook it from outboard balcony corridors.

Though inspired in large part by the client's emphasis on exposure to the outside landscape and the resulting abundance of daylight, as well as the demand for flexibility, the elegant lighting/partition system designed for CIGNA creates a handsome and polished interior landscape. Artificial and natural light combine to produce glare free illumination overall, while the moveable glass partitions open unbroken views to the atrium and the outdoors. Ceiling beams were dropped below the ceiling plane to a height of 9.5 feet and spanned by light tubes on six-foot centers. These not only provide indirect light but support the partitions.

The balconies connect inner office spaces with the atrium, thus allowing an abundance of natural light to aid in the illumination of the office spaces.

A four-story atrium provides the focal point of the CIGNA headquarters. Natural light in abundance and many plantings bring the outdoors inside.

The outboard balcony corridors are heavily used and they allow employees full view of the atrium throughout the day.

Natural light brilliantly illuminates the atrium and balconies.

A pool and plantings give the atrium an outdoors feeling.

National Westminster Bank

Project: National Westminster Bank Headquarters
Location: New York, NY
Lighting Designer: Wheel, Gezzstoff, Friedman
Interior Designer: Space Design Group
Photographers: Mark Ross, Elliot Fine

The National Westminster Bank Headquarters project, located on Water Street in the financial district of Manhattan, occupies 400,000 square feet on the 16th-31st floors, and the second and third floors. Long range plans call for full occupancy of the building.

The redesign of the space by Space Design included a comprehensive interiors program to accommodate: a retail bank branch and concierge station, service facilities, general open plan office spaces, sheltered workstations for clerical, secretarial, back-up and some mid-level managerial staff, private offices, conference rooms, executive offices and dining rooms, plus kitchens. Most of the major structural renovations occurred on the top floors, but special core treatments were applied to all.

Task/ambient lighting, consisting of evenly spaced single-tube fluorescents were positioned to remain constant in the general office space floors, regardless of furniture reconfigurations. Space dividers run perpendicular to perimeter partitions, that channel inter-office visibility from one end to the other. The dividers are wood, with clear glass panels. To also aid in defining space, various colors are used (primarily in carpeting) and half-round shapes, establishing a theme design motif.

In the executive offices, a staircase was added to connect the 31st and 32nd floors. The 31st floor is a dining room complex, consisting of eight private dining rooms.

Detailing on the top two floors is of fine quality: marble, mahogany, maple, oak, American walnut and other woods are used, conveying an air of elegance and sophistication. The half-round shape theme is continued in ceiling treatments, flooring, and in area structures.

Waiting room as seen from the reception desk.

*Open stairway, one of three, between
every two levels.*

Shared work spaces near core wall.

Reception area of a typical general office floor, seen from elevator lobby through logo-marked glass doors. Curved shapes, marble and custom carpeting are design elements that are repeatedly applied throughout.

The 16th floor cafeteria.

One of the private dining rooms. Note the decorative lighting fixture centered above the table, continues the main color scheme of this eating area—yellow.

The stairway that connects the two executive levels. Chrome railings and bannisters nicely reflect the light from recessed ceiling fixtures.

General office area contains half-round fascia protrusions posing as canopies or topping glass-enclosed work spaces. The half-round theme is vividely shown in this area.

Work area extensions capped and basebound with half-rounds and filled with translucent glazing.

Wood burl columns add elegance to the executive level elevator lobby.

Conference/meeting lounge adjoining boardroom.

Reception desk and lounge seating framed with curved partitioning.

U.S. News & World Report

Project: U.S. News & World Report Corporate Headquarters

Location: Washington, DC

Lighting Designer: Lighting Design Collaborative, John C. Sarkioglu, Gary A. Garofalo, Philadelphia, PA

Interior Designer: Deupi & Associates, Teresita Deupi, Principal-In-Charge, Washington, DC

Photographer: Max Mackenzie

About 80 percent of this 200,000 square foot corporate headquarters project contains open plan, with the remainder dedicated to private office space. In addition to the interior lighting design, Lighting Design Collaborative developed all the task and ambient lighting components along with the electrification design for the Design Group furniture system used throughout the space. Additionally, Lighting Design Collaborative coordinated with Underwriters' Laboratories to have all the designed components photometrically tested and approved. Employee workstations are comprised of wood systems furniture using work surfaces, partial height panels, overhead storage and files.

Since U.S. News & World Report is a highly automated magazine, a majority of workstations contain video display terminals (VDT's) connected to computer systems, which have virtually taken the place of typewriters. The high density of VDT's with their reflective, mirror-like screens, is a major consideration in the design of the lighting system. In response to this, a task ambient lighting system is integrated into the furniture system and architectural elements. The ambient lighting consists of indirect fluorescent fixtures mounted within the top of the workstation storage cabinets. This approach reduces the possibility of the reflection of bright ceiling fixtures on to the VDT screens which can washout the printed text displayed on the screen. To supplement the ambient lighting, fluorescent task lights are integrated into the bottom of the storage cabinets or on the desk top to provide higher levels of illumination on the work surfaces.

Round or rectangular direct/indirect fluorescent pendants are used over free standing desks or tables to provide ambient as well as task lighting.

In conference rooms, direct/indirect fluorescent pendants are also used over tables in conjunction with recessed fluorescent ceiling coves used to illuminate presentation walls above credenzas.

To maintain the lighting theme in corridors, indirect mercury vapor uplights and backlighted planters are built into firred out architectural columns and walls. Major artwork walls are illuminated with recessed incandescent wallwashers.

In the two-story library, a similar task/ambient approach is incorporated into the workstations and carrels located in the space. The bookshelves located around the perimeter of the space are lighted by a continuous fluorescent light slot recessed into the ceiling. (Since the glass walls of the library open the space up to the adjacent office space, the lighting treatment is similar and in the same context as the office lighting, but still honors the particular functions and architectural characteristics of the library.)

Flanking each elevator lobby is a curved reception space and curved display niche. Recessed light slots in the ceiling of the lobby and reception area reinforces the architectural geometry and provides general lighting. A cold cathode source is used in the slots to accommodate the curved design. Recessed incandescent coves provide a warm backdrop in the display niches, while recessed adjustable incandescent accent lights highlight three-dimensional objects that are on display.

About 80 percent of this 200,000 square foot corporate headquarters project contains open plan, with the remainder dedicated to private office space.

This reception area is illuminated by direct/indirect fluorescent pendants. The piece of artwork is illuminated with a recessed incandescent wallwasher.

Flanking each elevator lobby is a curved display niche. Recessed light slots in the ceiling of the lobby and reception area reinforces the architectural geometry and provides general lighting.

Major artwork walls are illuminated with recessed incandescent wallwashers.

A continuous fluorescent light slot recessed into the ceiling illuminates the news room.

Direct/indirect fluorescent pendants are used over free standing tables to provide ambient as well as task lighting.

In the two-story library, the book shelves located around the perimeter of the space are lighted by a continuous fluorescent light slot recessed into the ceiling (similar to the lighting treatment used throughout the remainder of the office space).

*In this office, a recessed fluorescent
ceiling cove is used above the credenza.*

To supplement ambient lighting, fluorescent task lights are integrated into the bottom of the storage cabinets or on the desk top.

Fluorescent task lights provide higher levels of illumination on work surfaces.

Commerce Bank

Project: Commerce Bank Corporate Headquarters
Location: Cherry Hill, NJ
Lighting Designer: John C. Sarkioglu, Gary A. Garofalo, Lighting Design Collaborative, Philadelphia, PA
Architect: Tarquini Organization, Camden, NJ
Interior Designer: InterArch, Inc., Shirley Hill, Principal-In-Charge, Cherry Hill, NJ
Photographer: Jack Neith, JDN Photography

Commerce Bank's 25,000 square foot corporate headquarters facility consists of a 10,000 square foot main banking floor with the remaining space being dedicated to executive office and conference areas for this New Jersey-based bank.

The building features an atrium with a myriad of focal points including the Bank's transaction and platform areas along with lush 25-foot trees, water fountains and free-standing sculpture. Ambient illumination for the atrium is accomplished through indirect lighting equipment concealed behind plant material in planters which organize the mezzanine balcony. The general lighting theme is theatrical in attitude and illuminates the various focal objects through a mix of downlights and accent lights creating predetermined modeling shadows. Functional fluorescent equipment is integrated and detailed into customer check writing and teller counters while stairs lead to the executive office floor above.

Adjacent to the atrium is the elevator lobby, in which a two-storey water-wall feature is situated with cascading water terminating behind a planter arrangement. The water-wall is illuminated by a series of incandescent lamps located in the spandrel above. This lighting approach grazes the marble wall behind the cascading water, as well as highlighting the planter below and increasing the specularity of the water as it strikes the basin.

In the executive suite, all offices are glass enclosed with views into the lobby corridor. Recognizing the aspect of borrowed light from the private offices, the ambient illumination was supplemented with ornamental fixtures featured about polished columns. The ornamental fixtures, designed by Lighting Design Collaborative had designed the firm's original task/ambient lighting and electrical componentry. The enclosed offices also feature a custom designed round wooden pendant fixture that was inspired by the round bullnose and finish of the furniture. Its task/ambient performance is in keeping with the vocabulary of the furniture as well as providing for a glare-free glass corridor wall. The artwall is punctuated by recessed incandescent wallwashers to balance the room.

The President's office features a ceiling coffer indirectly illuminated via two wooden millwork beams containing a fluorescent source. The marble walls are revealed with an incandescent bath of light, the seating area is illuminated with downlights and the airplane sculpture highlighted with adjustable incandescent accent lights.

The main boardroom is washed in incandescent light via a linear downlight cove and the table is articulated by a light slot in the ceiling that also provides task lighting. The table itself is illuminated with three custom designed etched glass pendants with the corresponding lamps being contained in the ceiling. The end walls are lighted with an incandescent recessed light slot to highlight artwork and wood surface. The flower arrangement along the marble wall is accented by incandescent downlights.

Functional fluorescent equipment is integrated into customer areas.

The water wall (adjacent to the lobby atrium) is illuminated by a series of incandescent lamps located in the spandrel above.

Ambient illumination for the atrium is accomplished through indirect lighting equipment concealed behind plant material in planters which organize the mezzanine balcony.

In the executive offices downlights are also used to highlight the floor finishes.

The secretary areas feature Geiger
International furniture systems, for
which Lighting Design Collaborative had
designed the firm's original task/ambient
lighting and electrical componentry.

The general lighting theme is theatrical
in attitude and illuminates the various
objects through a mix of downlights and
accent lights.

The table in the main boardroom is illuminated with three custom designed etched glass pendants with the corresponding lamps being contained in the ceiling.

In the executive suite, all offices are glass enclosed in with views of the lobby corridor.

Ornamental fixtures are made of wood and contain fluorescent equipment that uplights the area and gives the illusion of floating around the polished chrome columns.

The artwall in the executive offices is punctuated by recessed incandescent wall washers to balance the room.

Ornamental fixtures give the illusion of floating around the polished chrome columns.

The lobby corridor uses ornamental fixtures to add ambient light.

Glare free glass corridor wall works well with the lighting in the executive offices.

Norstar

Project: Norstar Bank
Location: New York, NY
Lighting Designer: William Lamb Agency
Architect: The Architects Collaborative
Interior Designer: The Architects Collaborative
Photographer: William Lamb Agency

This adaptive reuse of Albany's historic Union Station involved the design of lighting both for rehabilitated historic spaces and for newly created areas of contemporary workspace, often in close proximity. The particular challenge was to provide lighting appropriate and comfortable for a modern purpose while retaining the character and ambience of a turn-of-the-century train station. Recognizing that bright surfaces attract the eye, the design seeks to use architectural features and other objects of interest as the principal sources of light, or at least as the principal apparent sources of light. As far as possible, the lighting is treated as an integrated element of the architectural design, clarifying and reinforcing the organization and features of the building rather than distracting from them.

In the main lobby, for example, the lighting mixes recessed low-brightness sources with decorative fixtures sympathetic to the style of the old hall. The richly coffered ceiling, one of the major architectural features, becomes the main light source for the spaces below, uplighted from the new custom incandescent chandeliers. These come in two sizes, scaled to their respective coffers, and designed and positioned carefully to maintain a defining contrast between the bright coffers and the darker dividing beams. As a complement to the ceiling, some of the other features below are highlighted. Narrow-beam metal-halide spotlights recessed in the small squares of the central coffers accent the trees and serve as grow-lights. Aimable narrow-beam incandescents, similarly recessed, accent other features such as the entrances and the reception desk.

The main lobby is flanked by working office and conference space on three levels, open to the lobby or separated from it by windows. The use of indirect fluorescent uplighting and low-brightness recessed downlights in these adjacent spaces minimizes any glare or distraction they might create when viewed from the lobby itself. This general strategy is carried into the newly-created offices with less historic character. Here, an attempt has been made to concentrate ducts and other mechanicals at the interior core, allowing a higher ceiling at the perimeter. This higher ceiling allows better penetration and distribution of

daylight from the exterior. The artificial lighting in these zones is provided by indirect fluorescent uplights, which blend with and supplement the daylight. The resulting clean and uniformly bright ceilings provide lighting which minimizes shadows and glare, particularly important where video display terminals are used extensively. Task lights supplement this ambient lighting where required. In the lower-ceilinged areas,

recessed downlights are used, supplemented by fluorescent wall-slots to help provide orientation at the cores.

In private executive offices, the owner sought an intimate, residential character. Here, unobtrusive fluorescent uplights balance natural light on the ceiling, recessed accent lights highlight book-shelves and artwork on the walls, and floor and table lamps provide local task light. Low-brightness 2X2 parabolic fluorescents are positioned over the desk for use when a higher level of working illumination is desired.

The main lobby with original ceiling and relocated cast-iron facades. New custom-designed chandeliers uplight the coffers. The recessed metal-halide tree lights are virtually invisible from normal angles of view.

Working office and conference spaces flank the lobby. Careful control and placement of light sources in these spaces prevents glare and distraction from the features of the lobby.

Corporate
Offices

Second floor reception area at the end of the main lobby. Incandescent accent lights concealed above highlight features like the reception desk and corporate logo, and supplement daylight from the restored skylights when required.

The large lobby chandelier, with recessed metal halide accent lights in the small squares at the corners.

Open office area with raised ceiling at the perimeter. Indirect fluorescent uplight blends with and supplements daylight on the ceiling surface. Lower ceiling with recessed downlights defines circulation area.

Indirect uplighting provides even, glare-free illumination in open office areas, reducing veiling reflections on video display screens. Task lights are integrated with the furniture system.

Ingalls

Project:	Ingalls, Quinn & Johnson
Location:	Boston, MA
Lighting Designer:	The Architects Collaborative (TAC) Cambridge, MA
Architect:	The Architects Collaborative
Interior Designer:	The Architects Collaborative
Photographer:	Nick Wheeler

It was important that the building blend carefully with the scale and architectural character on either side: 19th century brownstones and the large scale Prudential Center directly across the street. Massing setbacks reduce the apparent size of the building, so that it blends with adjacent mid-rise townhouses and minimize shadows on nearby open spaces. A series of bowfronts along the street and the exterior of patterned brick, granite and cast stone is a graceful blending of the old with the new.

The 72,000 square foot headquarters of Ingalls, Quinn & Johnson, a 400-person advertising and public relations agency, occupies the upper six floors of an eleven-story building that the agency co-developed with a Boston real estate concern. A division of one of the advertising industry's international giants, with a diversity of blue chip accounts, Ingalls enjoys a highly visible position in the community and an excellent reputation for creativity and quality performance.

The agency wanted their headquarters to reflect this; to make a clear, appropriate image statement while fulfilling all functional and programmatic demands. Agency management was committed to providing the highly educated and motivated staff with working conditions that would support their professional 'pursuit of excellence' as well as promote their well-being overall. The client wanted the headquarters to provide a friendly, congenial atmosphere that would have a positive effect on staff attitude and allow for some individual expression in the workplace.

The interior plan, based on the patterns of work flow, communication and functional relationships within and between the agency's eight departments, is organized by a complete circulation route around the floor (except on floors ten and eleven, where there is a double height main reception area and balcony). Each floor is a mix of private and public spaces, with large and small offices and work spaces; the footprint is repeatable from floor to floor to optimize the movement of departments without the interruption and expense of construction. A projected growth rate of 30 percent in the first year of occupancy necessitated maximum flexibility in the plan, which had already increased by 12,000 square feet (an additional floor) in the middle of the design process, when the original client merged with another agency.

The tenth floor main reception area is a balconied, two-story space that acts as a focus of the headquarters image and establishes a sense of significance and arrival, conveying to clients and visitors alike the agency's strength and credibility. It doubles as a function area that can accommodate meetings of the entire staff and other large gatherings. This lobby design is an extension of the street level lobby of the building itself.

Each floor has a designated reception area and adjoining conference room, designed so that clients and other visitors cannot see directly from the reception space into any office or work area.

Private offices were required by approximately two-thirds of the staff. The building's variety of window shapes and sizes, including bowfronts and large square windows, were used to give character to the offices and other interior spaces; the space division also makes the most of available balconies and city views. In addition to maximum use of building perimeter, incorporation of natural light was a design priority.

In addition to a variety of conference facilities that include the main conference room with built-in AV systems and smaller conferences spaces throughout the plan, special conference areas include a variety of accessible exterior balconies for outdoor meetings, receptions and other formal and informal uses.

Custom lighting fixtures in both the building lobby and the main reception area establish the agency's design 'signature.' Symbolic of art, geometry and graphic quality, the linear black compositional element is carried into other furniture and accessories in forms, materials and detailing, including balcony rails. Other materials include black painted metal with polished chrome blades. The agency's visual statement is further reinforced by an art/graphics program.

The balcony formed by one of the massing setbacks allows a place for employees to go outside without having to leave the building.

Each floor has its own reception area and conference room. Note, the curved glass wall imparts an open feeling.

The main entrance of gray and black marble sets the tone for this building.

The employee cafeteria relies heavily on natural light.

A curved balcony and the use of plenty of natural daylight imparts on open feeling.

The main lobby.

The curved balcony in the main lobby adds architectural impact to the primary space in the building.

Signage, inconspicuously lighted, adds to the air of elegance.

Overhead lighting in the conference room adds inconspicuous and subtle illumination. Note tracklighting aimed at the walls for added illumination and esthetic effect.

The large windows in the conference room add a considerable amount of natural light to the room, as well as provide a breathtaking view of Boston.

A glass panel, lighted from the sides by fluorescent lighting is another form of signage used.

Ingalls, Quinn & Johnson

Design Group
Direct Response
Broadcast Production
Print Production
Marketing Services
Traffic

One of the conference rooms located
on each floor in the building. Recessed
overhead lighting is not necessary on
sunny days.

Vertical blinds on the conference room's glass wall can be closed to allow for privacy.

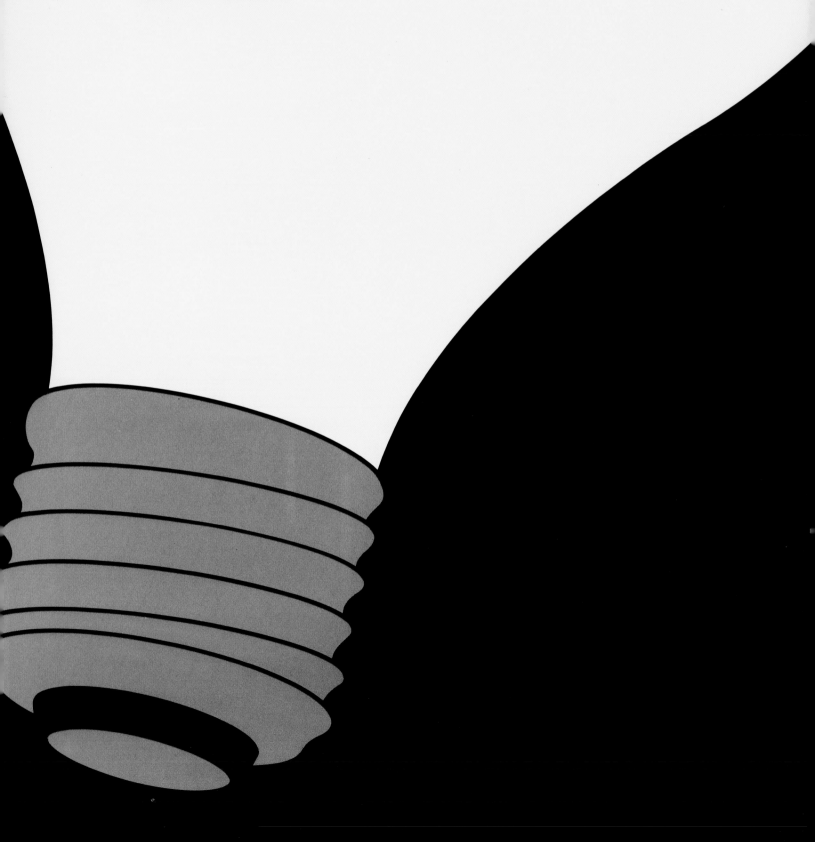

CHAPTER 2
SMALL PRIVATE OFFICES

Smaller companies may not have the need to utilize an entire building. Therefore, their offices must be distinctive and portray their company's personality with a much smaller amount of space to work with.

The lighting problems encountered in smaller offices are similar to those encountered in larger corporate headquarters. Ambient lighting problems have to be addressed, as well as task lighting problems, and ornamental (esthetic) lighting.

This chapter features a variety of offices —from the highly decorated and unique offices of Aesthetic Creations, to the functional offices of Dolby Labs.

Grey Advertising

Project:	Grey Advertising Presentation Center
Location:	New York, N.Y.
Lighting Consultants:	John C. Sarkioglu, Gary Garofalo, Lighting Design Collaborative
Interior Designer:	Joseph L. Rosen, Interior-Architectural Designer
Project Manager:	Nancy Preston
Designer:	Rosa Hurado Ritchie
Photographer:	Jaime Ardiles-ARCE

A small presentation area and screening room was designed with viewing comfort being foremost in intent. Through the dimming of incandescent pinhole downlights, a low level residential mood, in keeping with the plush comfort of the furniture, is provided. This serves to focus attention on the film or video being presented.

Executive offices are along the window wall and provide spectacular Manhattan views. Recessed incandescent downlights provide general illumination for desk and seating areas and matching recessed wallwashers illuminate the art walls.

This 60,000 square foot multi-media presentation and screening complex was designed with the intent of flexibility in lighting and audio-visual usage to allow complete freedom for multi-media techniques, as well as comfort for the viewing pleasure of the clients.

The reception rotunda consists of a circular space with a domed ceiling indirectly illuminated by a continuous incandescent light cove to help define the space. The reception desk and specular marble floor are highlighted through the use of pinhole downlights. Behind the receptionist, an incandescent accent fixture is intended to illuminate a future art selection.

Adjacent to the reception area is a small conference/lounge facility. The theme of small aperture incandescent pinhole downlights is carried out here while recessed incandescent fixtures softly bathe the granite walls.

The major presentation room reflects the intent of presentation flexibility. General lighting was provided by recessed pinhole downlights. These small aperture fixtures were specifically selected because they do not call attention to the light source itself, but rather subtly accent the presentation areas. The side fabric walls were envisioned as potential presentation walls, which resulted in the use of incandescent wallwashers to illuminate these areas. Low voltage accent fixtures are used to highlight items in the niche on the side wall, in this case a flower arrangement.

To further complement presentation techniques, LDC designed a low voltage preset dimming control system with the capability of four (4) preset scenes.

Incandescent pinhole lights are used in the conference/lounge facility.

The reception rotunda is indirectly illuminated by a continuous incandescent light cove to help define the space.

In the executive offices, recessed incandescent downlights provide general illumination.

Another view of the conference/lounge facility.

Low voltage accent fixtures are used to highlight items in niches on the side walls.

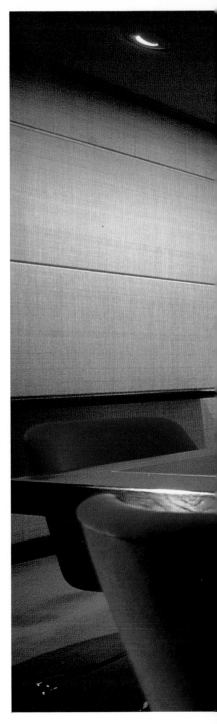

In the main presentation room, general lighting is provided by recessed pinhole downlights.

In this presentation room, recessed pinhole downlights are used for general illumination.

Dolby Laboratories

Project:	Dolby Laboratories
Location:	San Francisco, CA
Lighting Designer:	Luminae, Inc., James Benya, Principal-In-Charge
Architect:	Kevin Lemans, CRS Sirrine
Project Manager:	Jerry Gabriel
Photographer:	Douglas Salin
Linear Lighting Fixtures Manufacturer:	Peerless

The 65,000 square foot offices from the North American headquarters of the recently relocated Dolby Laboratories. The three-story building houses corporate executive offices, engineering research and development facilities, manufacturing areas, business and marketing offices, and a screening room especially designed to accommodate demonstrations of Dolby motion picture and professional sound equipment.

The building, built in the early 20th century, originally was used as an electrical distributor's shop and warehouse. After Dolby purchased the building, a series of renovations, reconstructions, earthquake reinforcements and systems improvements prepared the building for use as Dolby's national headquarters.

In the office areas, the exposed wood structure has been sandblasted and repaired to look clean and new. Ductwork and sprinkler lines have been coordinated carefully to create an organized appearance, without the usual camouflage of a suspended ceiling.

The suspended, six-inch-round, direct/indirect, lensed luminaires uplight the wood ceiling and provide general illumination as well. The fixture's high-efficiency uplight component furnishes approximately ten footcandles of indirect illumination, and the medium spread, soft-sidelight downlight provides 30 footcandles of direct illumination. The direct/indirect luminaire uses 3100 Kelvin, T-8 fluorescent lamps that operate with electronic ballasts. The total average maintained illuminance is 40 footcandles with 1.2 watts per square foot. An additional 35-100 footcandles is contributed by task lights in the workstations, depending on needed illumination.

To contain the noise level, an acoustical vaulted ceiling was installed in the cafeteria adjacent to the office areas.

The linear, horizontal theme is continued through the use of tubular fluorescent fixtures. In this case, however, the luminaires provide uplight only. Downlight and accent lighting are available from MR-16 adjustable fixtures that are incorporated into tubular housings. The 3100 Kelvin lamps reinforce a warm, cozy feeling in the space.

In the cafeteria, downlight and accent lighting are available from MR-16 adjustable fixtures incorporated in tubular housings.

©1987 Douglas Salin

In the adjacent office area, suspended six-inch-round, direct/indirect lensed luminaires uplight the wood ceiling and provide general illumination. Task lights are built into workstations to provide additional lighting.

©1987 Douglas Salin

Rogers Design Group

Project: The Rogers Office
Location: Coral Springs, FL
Interior Designer: Rogers Design Group
Lighting Designer: Rogers Design Group
Photographer: Michael Wall

In designing their offices, Lorraine Rogers and her staff aimed for the most pleasant, functional working environment they could achieve. Each of the eight rooms has subtle examples of the quality of work they do.

Working from just the shell, they first did the complete space planning and design, including work on the architecture and where they wanted the walls, electricity and lighting. The Rogers Design Group occupies 1200 square feet at 1750 University Drive in Coral Springs.

A wall-to-wall credenza, designed to be both functional and attractive, holds all the materials for presentations, with corner space for blueprints stored in individual tubes. A thick slab of glass serves as a desk and is held in place by a custom-designed pedestal in lacquered wood with a rainbow-colored top. On the wall behind the desk is an air brush painting, with all the colors of the rainbow blending into each other. A large, stucco textured stone urn is next to the desk.

The carpeting, in pale silver gray with pastel accents of mauve, violet and sea foam, provided the root of the color scheme. The wood trim is in pickled grey, providing for a warm, soft look throughout the offices.

Rogers' office adjoins the conference room, where the lower two-thirds of the walls are covered with ribbed carpeting, so designers can post their presentations. Above the carpeting are mirrors, giving the small room an illusion of endless space. The conference table and credenza are done in pickled tambour with soffit lighting overhead.

The conference room doubles as a showcase of the merchandise the design firm offers.

The accounting room, in shades of silver and gray with accents of lavender and fuchsia, is designed to be more utilitarian with built-in spaces for the computer terminal and other office machines.

The entranceway, with its vibrant colors, makes an instant impact when the visitor first walks in. The desk is a custom-designed, curved glass block. The front doors and a mirror are etched with the Rogers Design Group logo.

A corridor of glass block walls and mirrored ceilings leads to the studio, where junior and senior designers work. The hallway is lined with published articles about the firm. The reception area and corridor are tiled in blushed ceramic with coordinated blush grout. The walls are covered in a linen lacquered to match the tile.

The remaining spaces are custom-designed for records and storage. The firm also has an extensive resource library which is constantly being updated and enlarged.

Large windows provide an abundance of natural light in the office.

Mirrored ceilings in the hallway create an illusion of great height.

The reception area utilizes glass block
walls and etched glass to create an air of
elegance.

Skylights also provide additional illumination in the office spaces.

Aesthetic Creations

Project: The Sklarin Office
Location: Los Angeles, CA
Architect: Rieber O'Brian
Interior Designer: Roy Sklarin, Aesthetic Creations
Lighting Designer: Roy Sklarin, Aesthetic Creations
Photographer: David Zanzinger

A mirrored wall is a creative way to enlarge the size of the room.

The offices of interior designer Roy Sklarin and Aesthetic Creations, Inc., are located three blocks from the Design Center with ceiling heights of 15 feet. Sklarin designed the walls so they would be open to allow the skylights to filter the natural light.

In the reception area are 15-foot multi-height glass block walls with planters and etched glass doors opening into the conference room. Color scheme is ivory and black.

Marble is by Medisa with materials by the Tile and Marble Collection of Florida. The reception desk is an art deco original with etched glass top. Silk plants are from Make Beleaves.

The offices include full dining room and kitchen, conference room, drafting room, executive offices and two other offices.

In the reception area 15-foot glass block walls allow natural light into this area.

In the conference room, wall-mounted uplighted sconces add ambient light.

Deutsche Bank

Project:	Deutsche Bank
Location:	New York, NY
Interior Designer:	Interior Concepts
Lighting Designer:	Horton-Lees Lighting Design Inc.
Photographer:	Jaime Ardiles-Arce
Fixture Manufacturers:	Alkoo Mfg. Co., C.J. Lighting, Columbia Lighting, Contemporary Ceilings, Linear Lighting, Edison Price, Inc., Simes Co., Inc., Kurt Versen

In the open plan office area, care was taken to coordinate the desk layout to best take advantage of cross lighting created by the low brightness 1 foot x 4 foot parabolic fluorescent fixtures. The layout in conjunction with the "batwing" light distribution pattern of fixtures, resulted in illumination free of veiling reflections. The fixtures were run continuously in order to emphasize the strong organization and visual directionality desired by the interior designer. Simple coordination of the fixture layout with a glass acoustic partition results in the appearance of the continuous row running through the glass. Fluorescent wallwashers were located at the end walls to provide focus and visual definition to the room.

Fluorescent fixtures are run continuously to emphasize strong organization and visual direction in this open office layout.

Hill & Barlow

Project: Hill & Barlow
Location: Boston, MA
Architect: The Architects Collaborative, Howard F. Elkus, FAIA, Principal-In-Charge, Sherry T. Caplan, IBD, Principal-In-Charge, Robert Hicks, Project Manager
Interior Designers: David Howard, Tish Mullins

Hill & Barlow, one of Boston's most venerable law firms, is also growing rapidly. When they decided that they needed to move to a larger office, they selected two floors of International Place, a new financial district complex designed by Johnson/Burgee. The firm, whose clientele includes many architects and others in the real estate community, wanted an office that reflected their interest in design, and one that had the comfort and character of the eclectic, collegial Boston culture. The use of wood floors, oriental rugs, natural leather sofas and cherry finishes is a response to these objectives.

The building's geometry, a collection of forms arranged in a seemingly haphazard way in order to resemble a "village," demanded an unusual planning response. Each segment of the floor plate (circle, square, and rectangle) has its own column grid, and even the window bays vary on the curving perimeter from those on the straight walls, making a consistent application of typical hierarchical space standards difficult to achieve. The geometry also complicated mechanical and lighting planning. It became obvious that the sensible approach was to celebrate the opportunities presented by the building's geometry rather than to resist it.

The two-story atrium stair and lobby was the result of a desire not to have a single primary reception area which would favor the attorneys on that floor over those on the other, and to encourage easy communication within the firm. The point of collision of the circle and the square building forms seemed to demand a focal point be planned there, so that is where the lobby was placed. The lobby allows immediate views and orientation to the city and harbor. Similarly, the library is split between both floors and has a spiral stair connecting the floors. There is also a formal conference room on each floor. A roof garden was designed as an important amenity for the firm, and the dining room looks out onto it.

The client's program required a window office for each partner, but the high cost of space in the Boston market encouraged them to consider smaller offices than they had been accustomed to in their old quarters. A work wall/conference table arrangement to increase effective work area in less space was utilized, which also provided a good area for personal computers that are planned for each office. Partners' offices are arranged around the perimeter and have glass fronts. The full height glass fronts of the partner offices made it possible for the entire firm to share views and daylight. The attorneys required direct visual and vocal communications with their support staff and paralegals, so these staff members were placed directly outside each office. Associates were given the interior offices.

Table lamps are used in the reception
area to add warmth and to provide
needed lighting.

A spiral staircase connects the floors of this office.

Large windows provide natural light and a view to the offices.

Uplighted wall sconces add visual impact to the circular wall of windows.

The main dining room contains large windows to provide natural illumination and a view.

Andersen

Project:	Arthur Anderson Company
Location:	1 International Place, Boston, MA
Lighting Designer:	William Lamb Agency
Architect:	Architects Collaborative
Interior Designer:	Architects Collaborative
Photographer:	William Lamb Agency

Special characteristics of the client and the building helped shape this lighting design. Accounting activities to be carried on in the workspaces involved demanding visual tasks and extensive use of video display terminals. In addition, the client had an art collection to be displayed in these offices. The building plan itself, a square intersecting a circle, created a complex and unusual geometry.

In this context, the indirect ambient uplighting scheme chosen offered several advantages. The absence of recessed fixtures avoided awkward patterns on the irregularly shaped ceilings. The high quality of the indirect illumination helped reduce shadows, glare, and veiling reflections. And the uplights offered the opportunity for integration with a wall and partition concept which maintained the sense of openness and continuity in the space.

To avoid chopping the space into odd-shaped and potentially confining spaces, opaque walls and partitions were held down from the ceiling. Where private enclosures were desired, glass transoms closed the gap between walls and ceiling, allowing the sense of space to extend beyond partition boundaries. Fluorescent cove uplights concealed beneath these transoms provide not only uplight in the primary open areas but also borrowed light through the transoms to the enclosed offices beyond. Where the open office areas are large, this fluorescent uplight is supplemented with metal halide uplights integrated with the office landscape system. These uplights were custom designed to appear as a purposeful but unobtrusive part of the partition system, rather then as "stuck on" lighting fixtures, again contributing to the sense of continuity of irregularly shaped spaces. Fluorescent task lights integrated with the shelf system are provided for supplementary local illumination when desired.

At corridors and reception areas, full-height walls provide display space for the art collection. The lighting concept here was to allow the art, highlighted by recessed wallwashers, to provide all the light required for these areas. Incandescent sources were chosen for warmth combined with good color-rendering. In the private offices, the uplight scheme is continued with fluorescents concealed above wall mounted cabinetry, supplemented with adjustable accent lights to highlight plans and artwork and recessed downlights over the desks.

At the main entry corridor, recessed incandescent wallwashers highlight display space for the client's art collection.

Metal-halide uplights were custom-designed to be an integral and unobtrusive part of the partition system, located at the high corners. Similar forms can be used as planters, etc.

Filing and reception area uplighted by fluorescent coves beneath the transoms, which enhance the sense of continuity of the space and allow borrowed light to private offices beyond.

Conference room, with controlled beam incandescent downlights highlighting the conference table, and perimeter fill light provided by spill from the artwork.

Another view of the conference room, showing niche for display.

Ambient uplighting from fluorescents at edges (concealed here behind files) supplemented by metal-halide fixtures in the raised portion of the partition system.

Private office with glass transom, uplight over wall cabinets, and recessed adjustable accent light for artwork.

Private office at left benefits from
borrowed light through glass transom
over files.

Goldberg Law Office

Project:	The Goldberg Office
Location:	Fort Lauderdale, FL
Architect:	Donald E. Bryan
Interior Designer:	Roy Sklarin, Aesthetic Creations
Lighting Designer:	Roy Sklarin, Aesthetic Creations
Photographer:	Karl Francetic

Marble, black glass, and planned green areas are the keystones used to transform a former retail furniture outlet into the expanded offices of Fort Lauderdale's prestigious law firm of Goldberg, Young & Borkson, P.A.

The firm commissioned Roy Sklarin of Roy F. Sklarin Interiors of Fort Lauderdale to completely renovate its new home at 1630 North Federal Highway in Fort Lauderdale.

Holland Builders, general contractor for the project, worked closely with Sklarin, who prepared the artistic floor plans and elevations, and Donald E. Bryan of Donald E. Bryan Architects, Inc., consulting architect for structural, mechanical, and engineering work.

The single-story building with approximately 11,400 square feet of space was designed with clean horizontal and vertical shapes. A band of glossy black metal around the entire exterior of the building is softened by rounded corners. Because the interior is almost square, we used a core concept to maximize efficient use of space.

The interior core also has a feeling of the outdoors, however. Large, mature plants and trees provide a tropical back-ground around which Sklarin placed bands of functional space.

All of the multi-use spaces are in the center, the designer said, with a corridor around the central core. There is an inner ring for the secretarial staff. The attorneys' offices are in the outer band.

The inner core houses three different conference rooms, bookkeeping and word processing facilities and the firm's law library.

The main conference room, at the center of the building, is surrounded by planted areas. The plush reception area opens directly onto a planted area opening into the main conference room where the front wall is entirely of glass

with the rear planted area surrounded by mirror to enhance the feeling of the outdoors.

Two additional conference rooms are located on either side of the main conference room.

A planted buffer 54 inches high separates secretarial staff and attorneys from the corridors. A lounge and kitchen area are located in the rear of the building to serve the needs of office personnel.

Marble floors in the reception area are complemented by architectural carpeting with a mini-pattern. Oak desks are a feature in each office and interior walls are covered in fabric to soften the decor and provide increased noise control.

Every attention has been paid to details often missed in many commercial spaces. The placement of every light fixture assures the most natural, non-glare lighting. Even the air conditioning outlets and fluorescent lighting frames have been custom painted to match and blend in with the ceilings and walls.

The signage matches the design of the building. It is free standing and made of concrete with matching black aluminum bands and lettering to extend the exterior design of the structure to the entry as a welcome.

Ambient lighting in recessed portions of the ceiling create a visual focal point in the room.

Table lamps provide additional lighting in the reception area.

The unusual ceiling fixture in the conference room provides adequate task lighting.

Office alcoves are illuminated via fluorescent fixtures that illuminate the hallways as well as each individual cubicle.

The offices are simple, yet elegant.

SITE

Project:	S.I.T.E.
Location:	New York, NY
Lighting Designer:	Diane Berrian Viola, IALD, IES
Architect:	S.I.T.E.
Interior Designer:	S.I.T.E.
Photographer:	Peter Aaron, Esto

An industrial loft building by Louis Sullivan on the lower East Side of Manhattan sets the scene for the renovation of the entire second floor into new office space for SITE. The landmark building has recently undergone extensive restoration of the sculptural ornamentation of its facade. Special care has also been taken to recapture the simple elegance of the entrance lobby.

The second floor is the most interesting floor in the building. With a ceiling height close to thirteen feet, the 8,000 square foot space is punctuated by 21 full height columns. The Sullivan-ornamented capitals were completely restored by SITE and become a focal point and an intrinsic part of the office design.

A large entrance gallery also serves as the reception area. Consistent with SITE's unique philosophy of ''un-building'', 8'-0'' high partitions are made of unplastered, metal plastering-lath, providing semi-transparent screens that define and separate activities. The combination of these lacy, horizontal screens, the strong vertical columns and monochromatic white surfaces, laid the basic architectural concept to begin illuminating.

The conceptual lighting design was critical to successfully integrate with SITE's strong design style and historic nature of this Louis Sullivan building. A sensitive and light-handed approach was deemed necessary to bring out the interesting and complex architectural features in a quiet manner. The absence of a hung ceiling with all sprinkler pipes and HVAC systems exposed, further restricted and complicated possible locations for lighting fixtures.

Working with the partition layout, we recommended inverting two-lamp fluorescent channels into the tops of the partitions for uplighting. This solution provides the general lighting scheme throughout the space. Locations and intensities of lighting fixtures vary, according to the specific needs of different areas. For instance, the gallery is wrapped with a line of perimeter uplighting to balance the strong, direct sunlight from the south window wall. Louvers are used to soften the light where partitions abut full height walls.

Certain partition walls throughout the space were intended to display artwork. For this purpose, surface mounted incandescent track lighting was used to softly wash vertical surfaces were required.

The fluorescent uplighting brings the office to life, while providing a comfortable level of light for most tasks. Capital tops are highlighted, as requested by the designers. The addition of track lighting brings out another dimension of vertical emphasis. The most satisfying result is the combination of artificial and natural light casting subtle shades and shadows through the partition screens in an ever changing process, according to your position in the space. A single partition appears more translucent than looking through two or more partitions.

Lamp color was carefully selected to avoid adding a tint or greying the white surfaces. White fluorescent lamps with phosphors were chosen, as opposed to ultralume, cool or warm white. Quartz incandescent PAR lamps were chosen for their whiter color to blend with the fluorescents.

Lighting control provides additional flexibility, by using more or less uplighting and/or wall washing for entertainment functions.

This beautiful Louis Sullivan building is located on the lower east side of Manhattan.

The two-lamp fluorescent channels inverted into the tops of the partitions are the general lighting scheme for this office.

A large entrance gallery also serves as the reception area.

Two-lamp fluorescent channels are inverted into the tops of the partitions for uplighting.

Lacy, horizontal screens section off areas of the reception/gallery area.

Some of the partition walls are used to display artwork. Surface mounted incandescent track lighting is used.

Michaud Cooley Erickson

Project: Michaud Cooley Erickson
Location: Minneapolis, MN
Lighting Engineer: Monty Talbert
Architect: Radius Design
Designer: Marlea Gilbert
Photographer: Bob Witte

Leslie Davis of Michaud, Cooley, Erickson & Associates, Minneapolis, Consulting Engineers, took on the task of giving the firm's new offices an updated appearance and a pleasant working environment. Also desired were quality lighting for design and drafting tasks, a conservative high-tech look utilizing tenant package luminaires, and adherence to the Minnesota energy guidelines.

In the reception area, custom indirect wedges hide forward throw reflectors with 400-W T3Q/CL/SS lamps. Sawtooth metal shields provide decorative fanlike distribution.

In conference rooms, excellent color with 60 footcandles on work surfaces, and full range control are achieved. Adjustable downlights provide tight beam accent lighting on the tackboards.

Cabinets conceal luminaires in the executive offices. Quartz sources bring out the rich mahogany of the cabinetry.

In the design and drafting open offices tasks include reading, writing reports, drafting, and computer analysis. Building standard, tenant package, 1 x 4 parabolics inverted in custom housings provide ambient light at very low cost. Slots in the bottom of the housing allow for heat extraction and provide spill light on the bookcases. Undershelf fluorescent task lighting gives 100 footcandles on work surfaces. Adjustable task lights with 22-W circline fluorescent and 60-W incandescent combination mounted over the drafting surfaces provide 130 footcandles.

Adjustable downlights provide beam accent lighting.

Glass panels around this doorway provide impact.

In the design and drafting open spaces, undershelf fluorescent task lighting gives additional illumination to workspaces.

Fluorescent lighting provides general illumination in the computer areas.

Forsyth & Connors Advertising Agency

Project:	The Forsythe Office
Location:	Fort Lauderdale, FL
Interior Designer:	Rogers Design Group
Lighting Designer:	Rogers Design Group
Photographer:	Michael Wall

After concentrated conferences with the partners of Forsyth & Connors, Ms. Rogers designed all of the furnishings in the office complex to suit their individual needs and their preferences in decor. They are happy with the results.

In the private office of the more conservative partner, Ms. Rogers designed a replica of an English sitting room, with its Chippendale chairs and oak crown molding, and within the office of "Super Slick," the designer's name for the less conservative partner, she achieved a very dramatic effect with lots of chrome and an unusual desk.

The glass-top desk, which juts from the subdued gray wall at a 45-degree angle, was designed by Ms. Rogers, using triangular slabs of coral fossil stone stacked one layer on top of another to form the base.

The reception room contains a gray wall covering of textured suede fabric while the floor is glossy black ceramic tile. Two of the walls are mirrored, with the company's logo etched into the mirror directly behind the desk.

The high-gloss black desk top rests on a pedestal of eight-inch square blocks and sets at a 45-degree angle. Black lacquered linen drum tables were set in front of the wraparound gray chenille sofa, which features Chinese red and ebony silk pillows.

The conference room contains a 10-foot long, nine-foot high wall unit designed by Ms. Rogers to house all of the components needed to make an advertising presentation. Chrome tambour doors slide in and around the unit to conceal a wet bar, television monitor, and several video and audio components.

Subtle ambient lighting creates a pleasing mood in the reception area.

FORSYTH & CONNORS

An etched mirror creates a feeling of roominess in the reception area.

The unusual 45-degree angle of the desk
and the use of rock to form the base
creates a visual focal point of this office.

This office is designed to resemble a
chippendale style sitting room.

Another view of the conference room.

Fluorescent pendant lights provide ambient lighting in the conference room.

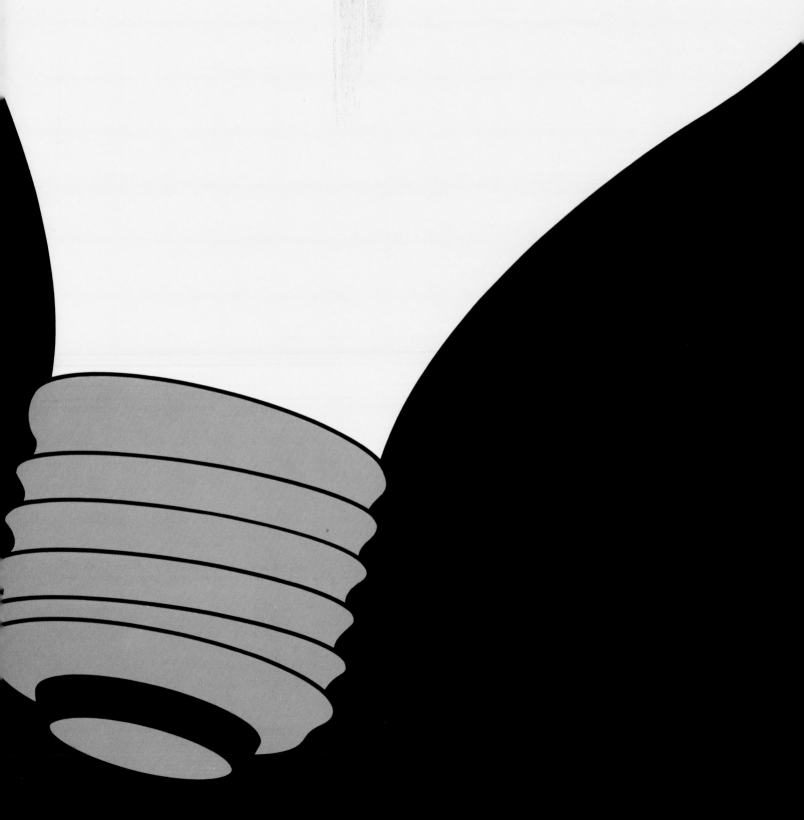

CHAPTER 3
DAYLIGHTING

n many offices, a frequently asked question is "what's the weather outside?" Many employees work in environments where windows, skylights, atriums and outdoor areas are nonexistent. The only light source for these workers is the fluorescent fixture above their desks.

Today, many lighting designers are incorporating daylight into their lighting designs. This chapter highlights those projects that use daylight effectively and efficiently with other lighting systsms.

Included is the ingenious atrium design of Thresher Square in Minneapolis, as well as the spectacular skylit offices of the famous design firm Vignelli Associates in New York City.

Elite

Project:	Elite Modeling Agency
Location:	New York, NY
Lighting Designer:	Haverson Rockwell
Interior Designer:	Haverson Rockwell
Architect:	Haverson Rockwell
Photographer:	Mark Ross
Lighting Manufacturer:	LiteLab

Haverson/Rockwell Architects provided Elite Model Management with a unique and functional design for their new headquarters building in a New York City four-story building. Natural light, introduced into both the ground floor and the second floor by separate sky-lights, gives the building's interior an airy, open feeling. The effect continues throughout the space with transparent windows that allow the natural light to illuminate spaces without a direct natural light source. The skylight effect is again simulated in other public areas with the use of cove lighting.

Track lighting was used in all of the offices and open booking areas with additional recessed lighting added in the executive offices. In addition, accent lighting was provided throughout by spotlighting artwork and plants, as well as emphasizing important architectural features.

A series of skylights allows an abundance of natural light into the agency.

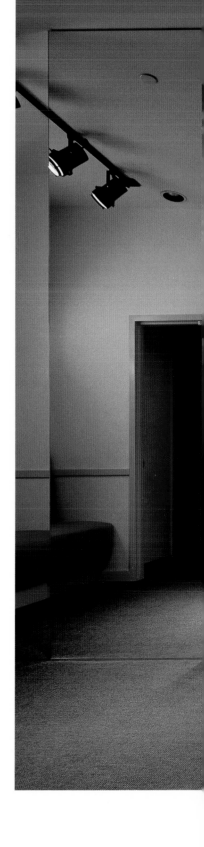

Incandescent ceiling cove lighting provides architectural and visual interest in the reception area.

Windows in offices that open onto the main corridor allow sunlight from the skylights to enter these inner spaces.

Track lighting provides directed light for photographs and artwork mounted on side walls.

Vignelli

Project: Vignelli Associates
Location: New York, NY
Designers: Vignelli Associates, Lella and Massimo Vignelli—Principals-In-Charge; David Law, Michele Kolb, Lev Zeitlin, Robert Skolnik, Robert Traboscia, Briggs MacDonald— project team
Engineers: John Valerio (structural), William C. Rose (mechanical)
General Contractor: Vignelli Associates
Photographer: Luca Vignelli

Massimo and Lella Vignelli's new Manhattan office on Tenth Avenue is based on the grid system of design. Unusual use of materials: galvanized steel, brushed aluminum, sandblasted glass, gray lead sealed in beeswax, add up to a modernistic and unique look in this major design office.

The adage "less is more" is apparent in these beautifully designed, tranquil offices.

Unfinished steel furnishings convey an architectural feeling.

The Vignellis created uplighted wall
sconces by custom designing rolled-steel
brackets to accommodate standard
industrial fixtures.

Skylights illuminate a large portion of the Vignelli office. In the public area, skylights illuminate the reception desk.

Skylights illuminate a cruciform receptionist's area.

Skylights and track lighting provide
plenty of ambient/task lighting in the
conference room. The walls were spray
painted in poly-chromatic gray particles.

Lella Vignelli's office contains an
anteroom that is a series of contrasts:
hard versus soft, elegant versus austere.
The travertine torcheres were designed
by Vignelli Associates.

In the small conference room, skylights also provide key illumination.

Skylights and the unusual wall sconces
illuminate the 24-station studio.

The office library also contains a
skylight. Task lighting is provided
underneath bookshelves.

Steelcase

Project: Steelcase, Inc.
Distribution Center
Location: Grand Rapids, MI
Lighting Consultant: Gary Steffy, Jeff Brown,
Gary Steffy Lighting
Design,
Ann Arbor, MI
Architect: The WBDC Group,
Grand Rapids, MI
Interior Designer: Steelcase Design
Services, Grand Rapids,
MI
Photographer: Steelcase, Inc.,
Grand Rapids, MI
Lighting Manufacturers: Atelier International,
New York, NY; Linear
Lighting, Long Island,
NY; General Electric,
Cleveland, OH; Kurt
Versen, Westwood, NJ;
Sylvania, Danvers, MA;
Valley City Sign, Grand
Rapids, MI

To emphasize the open well feature between the two floors of this facility, 3500K neon was introduced against a sky blue field. Downlights and adjustable accents are used to sufficiently illuminate the reception area while providing subdued backdrop to the neon-lit well and the lighted glass block walls.

The glass block walls surround conference space behind the reception desk. The newer 45-watt halogen PAR 38 lamps allow an incandescent cove to be introduced without excessive heat and power load. This wall wash not only dramatizes the reception area, but contributes to the sense of space and place when in the conference room. A task-oriented pendent completes the lighting.

Private offices bordering the building perimeter are fitted with indirect lighting to not only balance daylight brightness during the majority of work hours, but to provide a distinct visual focus for the occupants of the nearby open plan areas. Artwork is highlighted for additional focus and drama. All fluorescent lamps are warm white, while incandescent lamps are 90-watt halogen PAR 38 floods.

Private offices bordering the building perimeter are fitted with indirect lighting to not only balance daylight brightnesses during the majority of work hours, but to provide a distinct visual focus for the occupants of the nearby open plan areas. Artwork is highlighted for additional focus and drama. All fluorescent lamps are warm white, while incandescent lamps are 90-watt halogen PAR-38 floods.

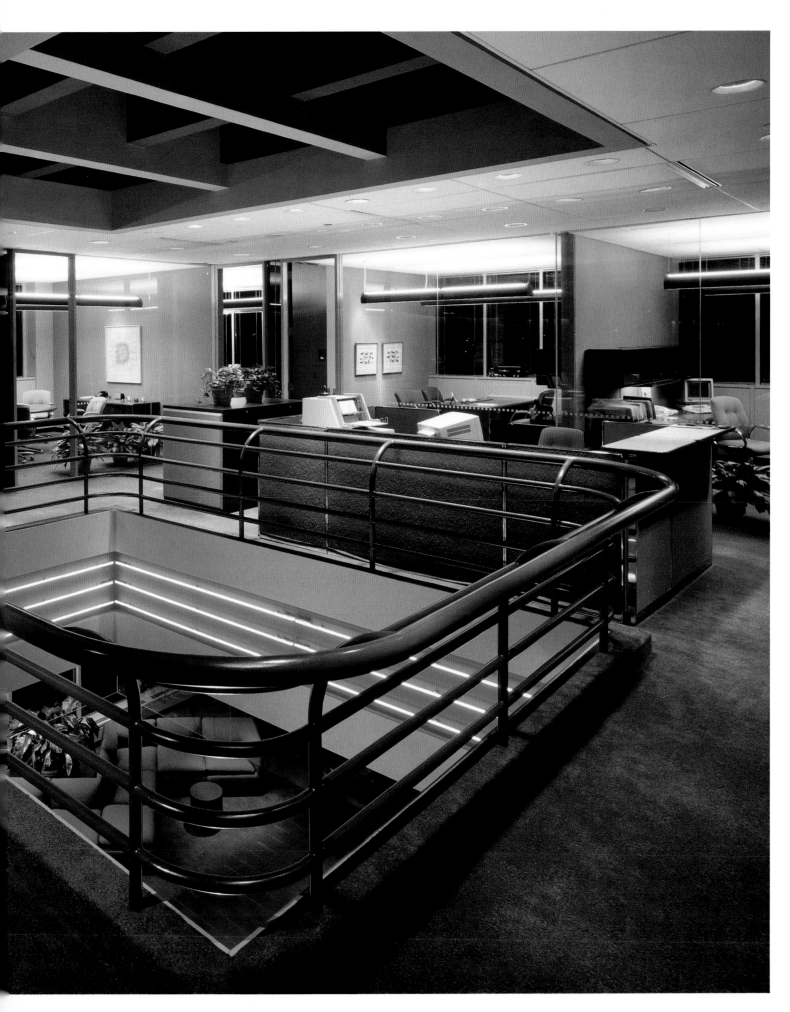

The glass block walls surround conference space behind the reception desk. The newer 45-watt halogen PAR 38 lamps allow an incandescent cove to be introduced without excessive heat and power load. This wall wash not only dramatizes the reception area, but contributes to the sense of space and place when in the conference room. A task-oriented pendant completes the lighting.

To emphasize the open well feature between the two floors of this facility, 3500K neon was introduced against a sky blue field. Downlights and adjustable accents are used to sufficiently illuminate the reception area while providing subdued backdrop to the neon-lit well and the lighted glass block walls.

Steelcase Western Division

Project: Steelcase, Inc. Western Division Headquarters

Location: Tustin, CA

Lighting Consultant: Gary Steffy, Jeff Brown, Gary Steffy Lighting Design, Ann Arbor, MI

Architect: The Austin Company, Irvine, CA

Interior Design: Steelcase Design Services, Grand Rapids, MI

Photographer: Steelcase, Inc., Grand Rapids, MI

Lighting Manufacturers: Archigraphics, Los Angeles, CA; Atelier International, New York, NY; Elliptipar, New Haven, CT; General Electric, Cleveland, OH; Halo, Elk Grove, IL; Intalite/Alcan, Charlotte, NC; Kurt Versen, Westwood, NJ; Lutron, Coopersburg, PA; Omega, Melville, NY; Peerless, Berkeley, CA; Steelcase, Grand Rapids, MI; Sylvania, Danvers, MA

Steelcase's offices and manufacturing facilities in Tustin, California provide a full-service office environment resource created for and dedicated to serving businesses in the Western states.

In 1958 the company opened its first manufacturing facility on the West Coast, and in 1964, after building a larger plant, established a division specifically to provide Steelcase products for the rapidly emerging markets of the West.

The Tustin facility was opened in 1972. At that time it consisted of 24,335 square feet of office and showroom space and a 420,000-square-foot manufacturing area.

Since then Steelcase has expanded the plant to nearly a million square feet and has created a total capability organization in the West, in order to meet the region's new and changing office environment needs.

On the Tustin project, the designers used the program to ensure that the lighting levels created by the furniture-integrated ambient fixtures would be sufficiently bright, visually comfortable, and of uniform intensity throughout the operations area.

The redesign involved two phases. Phase One involved constructing the addition. Then, in July of 1985, all employees were moved into the new space or relocated in temporary offices across the street. Phase Two, involving the renovation of the existing building, then began.

The original schedule called for the project to be completed by October, 1985. But, all renovations run the risk of unforeseen difficulties: the schedule was slowed by a series of problems with the ceiling and roof of the facility. A decision to install expanded underfloor deck for cable distribution in the existing building also took extra time, and final move-in was delayed until February of 1986.

Corporate support spaces, including this presentation room, maintain the indirect lighting theme established in the open and private offices. Here, neon in coves introduces a comfortable ceiling glow, while task-oriented incandescent downlights provide lighting for notetaking and reviewing product literature.

The private dining room has a series of neon coves for a comfortable dining ambience. For more drama, subtle color is introduced in the neon, with dramatic artwork and centerpiece highlighting to complete the setting.

The facility's entry not only functions as a reception and waiting area, but opens into an 'Action Area' where the latest Steelcase products are displayed. To highlight the entire display area and attract general attention, a series of red louvered pendant fluorescent luminaires is used. For product display, a second series of black tubular pendants houses incandescent track luminaires.

Architectural kiosks are introduced to provide visual interest and to help define circulation.

In private offices, pendant direct/indirect lighting equipment provides task lighting and a ceiling glow consistent with the open office indirect lighting. Fluorescent wall lighting is also used for visual continuity and to provide increased sense of spaciousness and pleasantness.

Fluorescent wall lighting provides a
sense of place and improved
pleasantness in the open office area. The
wall lighting also balances room
brightnesses, which helps with task
viewing concentration and with sense
of space.

The interior architects introduced window vignettes from one space to another for visual intent and to provide distant focus for eye muscle relaxation.

Thresher

Project: Thresher Square
Location: Minneapolis, MN
Interior Designer: BRW
Architect: BRW
Lighting Designer: BRW
Photographer: George Heinreich

The Thresher Building, located in Minneapolis, is an eighty-year-old building listed on the National Historic Register. It is a typical example of an historic warehouse building of an earlier era which has recently been converted to an office building as an adaptive re-use renovation. It has a masonry shell with a heavy timber interior structure and distinctive terra cotta work at its entries, on its window spandrels and wall surfaces, as was characteristic of some buildings of that period. The outside dimensions measure 180' x 122' placing it among the buildings of its type that when converted to office use, are too keep for a typical double-loaded central corridor, yet not deep enough to provide for a generous interior atrium. At six stories high, about 90 feet from first floor to reef deck, with floor-to-floor heights of between 14 to 16 feet, the building has an approximate gross square footage of nearly 10,000 square feet per floor in each of its two halves.

The original building was built in two stages. The west half, consisting of six floors, was completed in 1900. The east half, consisting of seven floors in the same overall building height, was completed in 1904. The renovation is also being implemented in two stages, beginning with the west half. All references, therefore, pertaining to areas and costs, apply only to the west half of the building unless specific reference is made to work—particularly exterior restoration and site construction—applicable to the whole building.

The principal stockholders of Bennett-Ringrose-Wolsfeld-Jarvis-Gardner, Inc. (BRW) are owners of Thresher Square, along with Hoyt Development Company. BRW, an interdisciplinary design firm, is also the architect of the project and Thresher Square's first and largest tenant. Hoyt Development Company is the first phase general contractor. Consultation to BRW on historic preservation and code approval was provided by Arvid Elness Architects, Minneapolis.

From the beginning, the architectural design effort has been directed at realizing the essential character of the historic building. Both the physical texture of heavy timber and brick and the spatial texture, represented by the interior rhythm of posts, beams and deck, prevails as the dominent aesthetic statement of the building. In order to achieve this the integrity of the interior structure has been left intact and its continuity made visible by creating a six-story atrium. While parts of the structure and deck were removed, the deletions were made in a way to dramatize the muscular capacity of the post and beam system and of the massive, five-inch thick timber deck. The atrium was created by selectively removing purlins (originally laid out 7'-4'' intervals to carry heavy equipment in storage), and allowing the deck to span 14'-8''. In some instances, removal of the purlins in a consistent pattern has resulted in cantilevering the deck as much as 5'-0''.

Making this happen without introducing an intervening supporting structure was accomplished by edging the deck at the atrium openings with a steel channel. The channel actually serves three purposes: first, it extends the structural capacity of the deck where required without introducing a readily apparent structural addition; secondly, it frames the combination of wood deck and three-inch lightweight concrete which constitutes the new floor; finally, it serves as a visual and physical base for the railing which completes the atrium edge.

Consistent with the effort to reveal and dramatize the structure of the historic interior was a careful consideration of the introduction of new systems—mechanical, electrical and architectural—necessary to convert the warehouse into a usable contemporary office building. Each new system received close attention in respect to the extent to which its introduction into the existing shell could be made without violating the overall legibility of the timber system. At the same time, each system had to be designed flexibly enough to respond to the demands of tenant requirements.

The mechanical system is a combination of a V.A.V. system for ventilation and cooling, and a hot water baseboard perimeter heating system. Although the interior brick faces of the exterior walls were never intended as a finished material in occupied space, the owners insisted on their remaining exposed, necessitating a substantial base-board system. The energy code is met by the addition of extra insulation on the roof. The ventilation/cooling distribution is through spiral ducts and V.A.V. boxes suspended from the deck above. The trunk ducts run around the atrium are set back far enough from the edge to fall behind the corridor demising walls on multi-tenant floors where such a wall would be required. Placing it there rather than at the exterior wall serves several purposes. First, it provides the greatest flexibility with the shortest trunk duct run; as tenant demands change, secondary distribution ducts may be run off it toward the exterior wall. These can be added to or subtracted from the system by "plugging" them into the trunk duct, or removing a distribution and blanking off the juncture, very much the way a plugmold electrical system allows similar flexibility. Secondly, the exterior windows are less likely to be visually cluttered or obstructed by the mechanical equipment if the trunk duct is not at the perimeter of the building. Finally, because the distribution pattern is visible from the interior as a coherent geometry and does not appear in bits and pieces (as in a perimeter system), its distinction from the building shell is clearly legible.

Electrical distribution is achieved with an under-carpet flat tape system which runs off a continuous buss-duct at the perimeter wall below the heating base-board. This allows both maximum accessibility and flexibility, and can accommodate systems such as CRT's and computers. Distribution can occur at any location along the buss duct. But total costs for the flat tape system tend to run higher as a result of the carpet tile required for accessibility. In its own space, BRW reduced the total square footage requirement for carpet tile by developing a decorative floor pattern that combines bands of carpet tile with broadloom panels. Although this necessitates that the electric tape to be run longer distances under the carpet tile bands in the pattern, it substantially reduces costs by minimizing the net amount of carpet tile required.

General lighting is supplied by fluorescent ceiling fixtures powered through a rigid conduit network concealed in the three-inch lightweight concrete leveling slab poured at each floor. At the atrium edges the underside of the exposed timber decks are uplit by fluorescent tubes mounted on the sides of the purlins and concealed by simple sheet metal shields.

The fire/life safety systems for the building represent a series of agreements worked out with Minneapolis Building Code Officials. To begin with, the building is entirely sprinklered. Since it was desired to keep the interior as open as possible, a smoke evacuation system was developed to minimize required enclosures. A huge duct in the basement carries fresh air to a structure in the center of the atrium at entry level. In the event of fire, air forced through the duct at high pressure creates a six-story air column which will draw smoke from occupied spaces at the perimeter into the center of the atrium. At the same time, the return air intake at the top of the atrium switches to full exhaust. This fresh air will flow from the perimeter of the building toward the atrium center, allowing occupants to get to the fire stairs in safety. Under this arrangement a portion of the building may be left entirely open to the atrium, without demising walls at the corridor, for whole-floor tenants. The amount of space which may be left open is determined by the capacity of the smoke evacuation system, the size of which was based on the owner's forecast of the proportion of single-tenant to multi-tenant floors.

The design of the architectural systems for the building follows three principles:

First of all, permanent improvements, such as railing and corridor walls reflect the original character of the historic structure, even if they were not actually elements required in its original construction. Wood and iron were selected as the materials. The introduction of dentil patterns in the railing cap and along atrium edges was inspired by similar detailing in the original ground floor exterior window frames. The decorative medallion motif, abstracted at elevator openings and elsewhere, appears in the terra cotta cartouches on the building's exterior.

Secondly, the spaces are left as visually open as possible. To accomplish this, the corridor demising walls, although deliberately designed with a massivity reflective of the building's historic character, are also designed to assure that the visual continuity of the space will always be unobstructed at the ceiling plane. To achieve this, a horizontal mullion is introduced into the demision wall at the door height of 8'-0''. Any privacy screening required by a tenant can be supplied by mini-blinds from this door height down. Above this level, the view is unobstructed. Tenants are encouraged to employ a similar approach to interior subdivisions.

Finally, where interior architectural elements have been introduced, their color and treatment are distinguishable from the historic framework of the building. Thus, even though the new architectural elements are historically derivative, they have been deliberately introduced in a way to maintain the clear legibility of the surrounding structure.

BRW's own space is an example of the application of all of these principles. Only a few enclosed spaces have been created; principally conference rooms and executive offices. The rest—with the exception of the photo studio and printing press, which had to be entirely sealed in sheetrock walls—employs an office landscape system. In the center of BRW's sixth floor space a glass walled conference room with its own metal roof floats above the atrium, visually detached from the surrounding building shell. Directly below it two offices also appear to be suspended and isolated above the atrium. These treatments, along with the removable heavy timber interior stair connecting the fifth and sixth floors, are not only expressive of the architectural concept of the building, but also serve BRW's need to achieve spatial and visual continuity among its many diverse activities.

Another design issue affecting the Thresher Square renovation was directed by the economics of the project toward maximizing the amount of lease space. The design problem was to develop a way to successfully transform the dark interior floor area at the center of the building into viable office space. Prior projects completed by BRW had demonstrated two important considerations which related to the Thresher building design issues. One was that it is light—particularly natural light—not view, which is the more critical ingredient in user satisfaction with office space. The other is that optically beamed sunlight makes it possible to project daylight into deep interior spaces.

In the course of developing an architecture of earth-sheltered buildings, two BRW architects, David Bennett and David Eijadi invented a system for tracking the sun's motion across the sky and directing a beam of sunlight into a deep interior space. It is called a Passive Solar Optics System (PSO) because it employs inexpensive panels of fixed reflective fresnel lenses to track the sun optically and deliver the light to a fixed area. It has proved to be not only an effective sunlighting system but a powerful general daylighting system on overcast days. The first application of the PSO system was in an experimental mode on the roof of the Civil/Mineral Engineering Building, completed in 1983, on the Minneapolis Campus of the University of Minnesota.

The PSO system, with its capability to cast sunlight at any predetermined angle, was an obvious candidate for solving the Thresher building's interior daylighting problem. It permitted the possibility of projecting natural light to the floor of a narrow atrium 100 feet below the roof. (Because the sun never exceeds an angle of 68° in Minneapolis, a conventional skylight would require an atrium of at least a width of forty feet for the sunlight to reach the floor for only an instant at solar apogee on June 22.)

The geometry of the atrium was developed based on the capability of the PSO sunlighting system and modified by other architectural design considerations. Among these was the desired width of the band of sunlight to light the atrium floor, its annual migration (determined by optical design) and its relationship to intermediate floors. The result is an L-shaped slot in floors three through six above a rectangular opening in the second floor. The south edges of the "L's" (on their long axis) and of the rectangle all align vertically. On the north edge of the "L's" each slot is successively wider, from top to bottom. The sequence is sixth floor: 12'-0'' feet, fifth floor: 14'-0'' feet, fourth floor: 16'-0'' feet and third floor, 18'-0'' feet. Because the second floor opening is a rectangle, the L-shaped upper floors also appear to be rectangular openings, with successively larger floor platforms projecting into them, rather than the L-shaped slots they actually are. This geometry, along with the penetration of sunlight (and bright daylight on overcast days) has transformed the normally undesirable interior floor areas of the Thresher Building into prime lease office space. At the same time, the narrow slot atriums have maximized the net leasable floor area, increasing the economic viability of the building from a net rentable of 78 percent to a net rentable of 83 percent.

The PSO sunlighting/daylighting system on the Thresher Building cost less to build and is a more reliable provider of light than a conventional skylight of the same size. A conventional skylight, both prefabricated and custom built was priced at about $50.00/sq. ft. The PSO skylight, including housing, glazing and lens material costs about the same amount. A conventional skylight, built as a clerestory, with a vertical glass wall is much less expensive than a sloped glass system, but would have contributed too little light to be worth constructing. A

glass-topped skylight would have been required, whether arched, sloped or flat, to deliver daylight 100 feet below. It could not deliver sunlight. All of these configurations have the disadvantage of being much more expensive than a vertical glass. Moreover, in Minnesota's frigid winters, a snowfall on even a steeply sloped glass surface will remain for days if not physically removed, despite the significant heat transfer through the glass from an interior space. The PSO system is glazed vertically like a cerestory but, because it bends light optically, it has the economy of the clerestory configuration while exceeding the light delivery capability of a glass-topped system.

The interior areas of most commercial buildings have no more access to sunlight and daylight than if they were deep underground. This may be as much as fifty percent of a moderate-sized building (say 80'-0'' x 80'-0''). The PSO system has application to all buildings in which the interior space can benefit economically and environmentally from daylight and sunlight. In respect to the renovation into office space of the ubiquitous, historic building-type represented by the Thresher Building, the introduction of the PSO system has the capability to transform the limited options provided by their dark interiors into opportunities for development.

©David J. Bennett, FAIA

Thresher Square is an 80-year-old warehouse that has been converted into office space in Minneapolis, Minnesota.

A cutaway illustration of Thresher Square.

A view of the skylights that allow the sunlight into the atrium.

Track lighting in individual offices enhances the sunlight from the atrium.

The atrium with a view of the levels of offices.

Another view of an office.

Glass walls allow sunlight into inner offices.

Glass walls section off the spaces into offices, yet keep an open, airy feeling.

Skylights bring natural lighting into darker spaces.

The sunlight highlights the natural brick of the atrium.

The atrium.

Wood beams evoke a feeling of a bygone era.

Natural sunlight bathes the atrium.

Clark Harris Tribble & Li

Project: Clark, Harris, Tribble & Li Architectural Office
Location: Charlotte, NC
Lighting Consultant: Francis Krahe, Francis Krahe and Associates, Newport Beach, CA
Architect: Clark, Harris, Tribble & Li
Interior Design: Clark, Harris, Tribble & Li
Photographers: Mario Carrieri, Milan Italy and Gordon Schenk, Charlotte, NC

This 44,000 square foot headquarters for an architectural firm is a renovated two-story warehouse and dispatch office for a trucking company. The non-descript, boxy building, originally built in 1947, had no exterior windows.

Given these conditions—and a very tight budget—the designers elected to create a lively, light filled building with a clear organizational system focused around a central atrium.

The skylit atrium allows natural light to spill deep into the building and is the central element around which the rest of the building's functions are arranged concentrically. Principals' offices and conference rooms border the atrium, screened by quarter barrel glass block walls; main circulation corridors surround these offices and conference rooms; architectural studios and other administrative areas are at the perimeter of this concentric system. Balconies overlooking the atrium on all four sides provide a sense of activity by allowing glimpses through circulation areas into the work areas. These same balconies allow a view of the skylight which provides a sense of connection with the outdoors.

The lighting system was designed to complement the character of light in the atrium while accentuating the architectural design and providing needed task illumination. In the day this system provides a bright work environment to reduce the contrast to the direct sun streaming into the atrium. At night the architectural features and details in and around the atrium are illuminated to reinforce the character and quality of the space.

From the atrium, views of the boardroom and principals' offices create an interesting sense of depth and space. The mulitiered ceiling vault of the boardroom is indirectly lighted, providing a dramatic backdrop and outlining a "Keyhole" window high in the wall above. Principals' offices are illuminated very pleasantly with general incandescent downlighting and a task luminaire over the desk, which provides some ambient light with a cleverly-placed reflective disk. The placement of the downlights was controlled to make the curved glass block walls "shimmer" with light when viewed from the atrium.

Continuous rows of 1 x 4 parabolic fluorescents provide glare free lighting in CAD areas.

Central atrium looking toward second floor boardroom.

The drafting areas are evenly illuminated with parabolic fluorescents placed in continuous rows 9 feet on center.

An anteroom links the boardroom and the atrium.

Fluorescent coves and downlighting reinforce the dramatic boardroom ceiling and balance the incoming daylight.

Principals' offices and conference rooms are behind curving glass block walls in each corner of the atrium.

Indirect fluorescent coves provide relief from the low ceilings of the reception area.

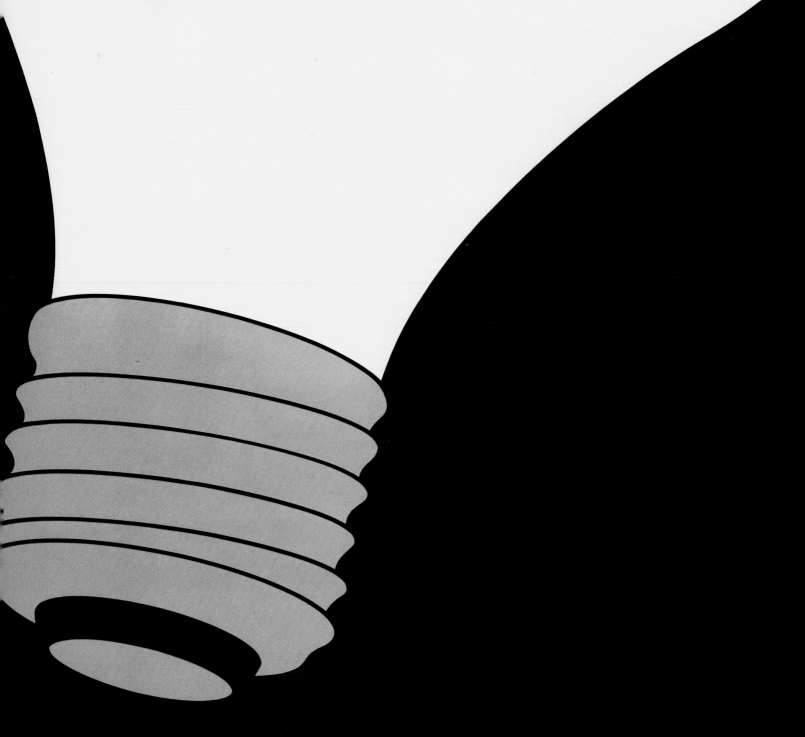

CHAPTER 4
RECEPTION AREAS, MEETING ROOMS AND EXECUTIVE SUITES

he lighting problems encountered in reception areas, meeting rooms, and executive suites are specialized in that illumination has to be provided to a large group of people. In addition, the lighting must be flexible enough to accommodate various lighting situations: meetings, seminars, presentations, audio-visual shows and many others.

This chapter presents projects that utilize ornamental lighting, ambient lighting and task lighting in various combinations to satisfy different situations—from the light and airy reception area of the Walker Dental Offices of Las Vegas, Nevada to the beautifully appointed conference facilities at Gulf State Utilities in Dallas, Texas.

Tower Records

Project: Tower Records Headquarters
Location: Sacramento, CA
Lighting Designer: Luminae, Inc., Janet Moyer, Principal-In-Charge
Architect: Robert Tanaka
Interior Designer: Patty Glickbarg
Photographer: Peter Marcus

©1986 Peter Marcus

©1986 Peter Marcus

Lyntas House

Project: Lyntas Advertising
Location: London, England
Lighting Designer: Sally Storey, Lighting Design Limited, London, England
Architect and Interior Designer: David Leon & Partners
Photographer: Bob Belton

The ceiling of the reception area and boardroom is studded with semi-recessed, low-voltage starlight luminaires that use 20-watt, 12-volt bipin lamps that provide general illumination. Low-voltage pinhole downlights have been placed over the reception desk and in the ceiling surrounding the columns to emphasize the columns.

The planters are highlighted with low-voltage directional luminaires. A preset dimming system balances the range of light sources to suit different times of the day. The interplay of light and shadow on the columns gives them dimension.

An attractive and pleasant space to be in, it is not cold, in spite of the color scheme that uses cool blue and glossy marble.

The ceiling in the reception area and boardroom beyond takes on a studded, starlike sky effect due to the semi-recessed, low-voltage luminaires.

Brooke Bond

Project: Brooke Bond PLC
Location: London, England
Lighting Designer: Sally Storey, Lighting Design Limited, London, England
Interior Designer: Charles Hammond
Photographer: Fritz Von Der Schulenburg

The boardroom is formal and traditional in its styling. General illumination comes from the ornate chandelier and table lamps, and tiny, tungsten, recessed downlights in the bays.

Cold cathode emphasizes the ornate ceiling and boosts the level of background light to 300 lux. Low-voltage miniature sources positioned over the table boost task light to 500 lux.

The delicate pinhole, low-voltage downlights do not intrude on the classic decor of the reception room—they provide general lighting. Directional sources highlight pictures and pathways.

Table lamps provide soft, local illumination.

In the dining room, daylight is softly diffused through muslin blinds and a patterned wood screen. Table lamps on the buffet balance daylight.

Additional illumination is provided by cold cathode when the room is used for board meetings. The low-voltage pinspot downlights highlight the center table display and provide task lighting when used as a boardroom. Three preset light settings are available to suit varied functions.

Cold cathode surrounds the perimeter of the room.

The dining room has three preset lighting settings to suit varied occasions.

Delicate pinhole, low-voltage fixtures provide general illumination without intruding upon the formal, traditional interior design.

Rich, wood paneling and ornate plaster ceiling are enhanced by illumination from recessed downlights and a crystal chandelier.

Standard Chartered Bank

Project: Standard Chartered Bank

Location: Bishopsgate, London, England

Lighting Designer: Sally Storey, Lighting Design Limited, London, England

Architect: Fitzroy Robinson Partnership

Interior Designer: Charles Hammond Limited

Photographer: Fritz Von Der Schulenburg

The octagonal shape and architectural detailing of the domed ceiling are echoed in the carpet pattern. Cold cathode lighting is used to delineate the raised dome and to indicate the entrance. The cold cathode runs along the perimeter.

General illumination is provided by silver cone downlights that give off a warm tungsten light that constrasts with and balances the cold cathode.

Framed photos and paintings are highlighted by low-voltage fixtures to add sparkle. The traditionally styled table lamps are a local, soft source of light.

The lighting system has been preset to furnish a variety of lighting moods throughout the day.

A focal point in the Director's Suite is the dome ceiling illuminated with cold cathode. Downlights illuminate the carpeting.

The rich materials are enhanced by the illumination—table lamps and small, recessed downlights.

A recessed cove of cold cathode runs along the perimeter of the entrance to the large, high-ceilinged suite. Downlights provide general illumination.

Akasaka Noa

Project:	Akasaka Noa Building
Location:	Tokyo, Japan
Lighting Designer:	Motoko Ishii, Motoki Ishii Lighting Design, Inc.
Architect:	Kunihide Oshinomi, Kajima Corporation Co., Ltd.
Photographer:	Satoshi Uchihara, Motoko Ishii Lighting Design, Inc.
Lighting Manufacturers:	ABC Trading Co., Tokyo, Japan

Akasaka, located at the heart of Tokyo, functions as a multiple commercial center of the metropolis. Bustling office buildings, hotels, and restaurants well represent the town's characteristics. This area is also within walking distance to the Japanese administrative center around the Diet Building.

The building stands on the busiest street of this area, called "Sotobori Dori." The facade of the building gives an outstanding view among the group of buildings there. Owned by a renowned Japanese securities firm, the first floor is used exclusively for counter services. From the second floor upwards, each floor serves as office space.

The lighting design of the first floor consists of the following:

The louver ceiling over the counter curves gracefully to follow the line of the counter. Pitches of louvers, as well as the arrangement of 40w fluorescent lamps are carefully calculated so as to avoid causing glare onto CRT's, and the digital display board of stock quotations. The surface of the ceiling itself is evenly illuminated. Indirect lighting by 20w fluorescent lamps is provided for the boundary part between the louver ceiling and the finished ceiling for a pleasant appearance.

Downlights at the regular ceiling are well arranged to coordinate with the location of air diffusers and the curved line of the wall. 100w metal halide lamps which are rarely used in office buildings are applied here for downlighting and furnish the space with a pleasant ambience.

The gently curved wall consists of a successive pattern of glass parts and columns. At the column tops, indirect lighting by 40w fluorescent lamps is provided to respond with the lighting design of the boundary part of the ceiling and helps to instill a pleasant feeling. At the top of the glass, 60w krypton lamps are installed to give adequate brightness onto the street outside for passersby at night.

The arrangement of 40w fluorescent lamps are carefully placed so as not to throw a glare onto the CRTs on the counter below.

Downlights in the regular ceiling are coordinated with the location of air diffusers and the curved wall, forming a symmetrical and pleasing arrangement.

Oracle Corporation / 20 Davis

Project: Oracle Corporation/20 Davis Drive

Location: Belmont Hills, CA

Lighting Designer: Luminae, Inc., James R. Benya, Principal-In-Charge

Interior Architect: Ehrlich Rominger, Laura Seccombe, Project Designer

Photographers: Douglas Salin and Steve Whittaker

The Oracle Corporation supplies the writing of computer software programs, primarily for advanced relational data-base systems. The offices have been designed to reflect the upscale, contemporary image of the successful, rapidly growing company.

The building that houses Oracle was originally designed to be a multi-tenant building, with a lobby intended to serve as a centerpiece for several tenants. Currently, there is only one other tenant —the 20 Davis cafeteria—in the building.

The main feature of the three-story lobby is a magnificent James Byrd porcelain enamel painting. In the daytime, the southwest-exposed glass wall provides considerable illumination. At night, concealed theatrical fresnel and ellipsoidal fixtures furnish color washes and auras on and around the piece. These effects were chosen by the artist in conjunction with the lighting designer.

General illumination comes from a double step coffer rimmed with 3000 Kelvin, 75 CRI fluorescent uplights. Two rows of recessed 175-watt metal halide atrium lights have been incorporated as adjustable plant lights. The exposed chrome-railed stairway is carefully illuminated with recessed slope-ceiling MR-16 downlights.

The building lobby (off the atrium) is conservatively lighted with a combination of coffer uplight and recessed, MR-16 adjustable accent lights. A pair of ornament uplights are used on each end wall beside small Byrd porcelain paintings.

The building's location in Belmont Hills makes it somewhat removed from commercial food services. The developer and Oracle insisted the cafeteria by a highly styled, attractive getaway and serve as an alternative to driving to a restaurant. The cafeteria features well-prepared, reasonably priced food.

The interior design theme is Italian-influenced and uses a sophisticated, modern color palette. The interior blends ceramic tile, laminates and fabrics in a light gray and charcoal base, with lavender, orange and alladon accents. Chrome and black leather seating serve as neutral elements.

A lavender glow is created by up-lighting the central coffer with fluorescent blacklight without the blue filter. The inner steps of the coffer are outlined in dimmable orange neon, which causes the coffer's halo of lavender uplight to dissolve in an electric array of lavender-orange overlays. White light is provided on the floor and tables by MR-16 downlights. Track fixtures, also using MR-16s, surround the coffer and are fitted with color filters to playfully lavish orange, alladon and lavender light upon the light gray divider laminates. As a finishing touch, six sleek Italian pendant fixtures are suspended over the banquette seating and are separately dimmed.

Since the Chairman of Oracle is a collector of fine Asian art and artifacts, he not only wanted a tasteful executive suite, but wanted a display created for some of his priceless collecton. The focus in the chairman's office is on five niches in the north wall. Each wall is lined with light strips to provide horizontal light and fill light, and with MR-11 attachments, key light. Incandescent luminaire rated at 2700 Kelvin provide horizon light. Art fill and back key light are accomplished with 3100 Kelvin MR-11 and 20-watt halogen bud inserts to the strip lights. Frontal key light is completed with recessed MR-16s in the ceiling, cooled to 3800 Kelvin. The reverse color temperature composition gives the whole wall a composed, two-dimensional appearance.

A variety of luminaires is used in the executive waiting area. Custom 3800 Kelvin cold cathode lighting in a radius coffer provides uplight. MR-16 fixtures with 3800 Kelvin filters are used to accent seating, art and circulation areas. Black paracube, 3500 Kelvin, T-8 lamps with electronic ballasts have been installed over the receptionist's desk. Wall sconces equipped with cool compact fluorescent lamps are mounted on the rear reception wall.

The boardroom coffer echoes the

waiting area coffer, with the addition of three beautiful pendant fixtures. Adjustable accent fixtures using MR-16s provide illumination for notetaking and highlight the display of two magnificent Chinese urns.

The chairman's office includes a work area, a seating area and an art wall. Ambient light for general work comes from a halogen arc torchiere. Dramatic accents are provided on all surfaces by recessed MR-16 accent lights. Credenza work light is provided by fluorescent 3100 Kelvin T-8 lamps mounted above a black paracube. The seating area is illuminated by MR-16 accent lights, balanced by the wallwashing of an antique screen on the side wall.

©1987 Douglas Salin

©1987 Douglas Salin

The lobby at 20 Davis features a magnificent James Byrd porcelain enamel painting. At night, concealed theatrical fresnel and ellipsoidal fixtures provide dramatic and colorful lighting.

The reception area of Oracle Corporation.

©1987 Steve Whittaker

Fine Asian art fills niches in the chairman's office.

In the boardroom, three beautiful pendant fixtures enhance the ambient lighting.

©1987 Steve Whittaker

In the employee cafeteria, a lavender glow is created by uplighting the central coffer with fluorescent backlight without the blue filter.

In the executive waiting area custom 3800m Kelvin cold cathode lighting in a radius coffer provides uplight.

AEtna Life / Financial Division

Project:	AEtna Financial Division Headquarters
Location:	Hartford, CT
Lighting Consultant:	Jeffrey Milham, IALD
Architect:	Jack Dollard, AEtna
Interior Designers:	Melvin Dwork and Claude Langwith
Photographer:	Peter Vitale
Contract Documents:	Russell Gibson von Dohlen

The Financial Division of AEtna Life & Casualty occupies 19 floors of the 37-floor City Place building designed by S.O.M. Chicago and built by former AEtna subsidiary Urban Development Corporation.

The 20th floor houses the chief financial officer of the corporation as well as the officers who report directly to him. Mr. Donald Conrad, the chief financial officer, and the other officers travel extensively to other major cities—New York, Chicago, London, Paris, Rome, Los Angeles, etc. It was their desire to have an executive office area in Hartford that would be comparable to the other offices that they visit in their buiness travels.

The design intent was to try to produce an interior that represented 150 years of good design in a cohesive but eclectic mix of traditional, modern, national and international furniture, fabrics, art and artifacts producing a timeless design.

The selection of furniture brings together modern pieces by Le Corbusier, Aalto, Saarinen, Ward Bennett and Mario Bellini with selective antique pieces, such as a rare Russian and Bessarabian carpets, a Chippendale bench and additional early 19th century English pieces, a Swedish folk cabinet, a Korean chest, and a Japanese lacquered hibachi—these pieces then again were combined with 17 contemporary pieces designed specifically for the 20th floor by the designers and produced by Cambium/L. Vaughn.

The fabrics range from French silk satin from Clarence House, needlepoint from Brunschwig & Fils to the newest lines from Knoll, Jack Lenor Larsen, and Gretchen Bellinger.

The art and artifacts include a diverse mix as well. American artist Christos, a tapestry by Stuart Davis, an 18th century Japanese screen, early American quilts, posters from the '20s and '30s, a rare set of 16th century botanical prints, contemporary photographs by Mark Golderman and Bert Stern, a recent work by Frank Faulkner, an assemblage by John Houck and a 16th century Japanese cedarwood carving of a Zuijin—guardians to the Emporer.

A black granite and taupe marble floor set the stage. Sand tones ranging from highly reflective to flat, combined with black lacquered core walls and columns, give this floor a strong architectural statement while providing a dramatic backdrop for this unique collection of furniture and art.

The Hartford firm of Russell Gibson and Von Dohlen with project coordinator Fred Saehrig produced the contract drawings and share the supervision of construction with the design team. A lighting consultation firm worked closely with the designers to provide a lighting system that reinforces the quality of the setting while providing suitable lighting for activities that are germane to the business that is conducted in this space.

Everyone, from the client to the designers, from the hard hats to the man who hand cut and fitted each marble and granite tile, formed a team who put their complete efforts to turn this floor from the ordinary to the refined.

The lighting in the slot and the reflection of the lighting on the floor create the illusion that was the designer's intent. . . a mysterious place of very beautiful and valuable materials, art and artifacts. The marble and granite floors actually continue into the adjacent rooms as seen here. The use of silver-bowl ellipsoidal reflector downlights with specular black cones gives a warm incandescent light to the rich materials used for the carpet and seating while minimizing the reflection of the units in the space.

A small seating area is directly adjacent to the wall for more informal meetings. The wall itself is totally black glass, wall to wall and floor to ceiling. All the audio-visual screens and VDT screens are handled by rear projection equipment in the adjacent room, but a mobile control center on a rolling cart allows coordination of the audio-visual presentations from any place in the conference room. Careful control of dimming the incandescent lighting and cold cathode lighting in conjunction with switching the fluorescents provides sufficient light in which to read and write while casting no reflections or light on the screen so that clarity of the images is excellent.

The wall with the Japanese wall hanging is the common wall with the reception area, but on this side, the wall-hung panels are wrapped in fabric. One low voltage accent light is used on each panel to highlight the fabric or to accent the artwork. The boat-shaped ceiling form over the table was determined by acoustical requirements, and silver-bowl ellipsoidal downlights are used to light the leather-topped table. The form of the ceiling is also treated as a floating object within the architecture by the use of a continuous cold cathode custom designed lighting slot on either side. The black wall in the background is the audio-visual wall.

Gulf States Utilities

Project: Gulf States Utilities Co. Corporate Headquarters
Location: Beaumont, TX
Interior Designer: Morris-Aubry Architects
Lighting Designer: Horton-Lees Lighting Design Inc.
Photographer: Charles McGrath
Fixture Manufacturers: Atelier International: Wall sconces; SPI Lighting: Cylinders, pendant fixtures

Lighting outside the boardroom washes the architects geometric forms in a diffuse way to naturally model them. This soft, subdued lighting relies upon the architecture to focus and frame the boardroom which, by contrast, is a brighter space with lighting elements which are themselves of visual interest.

The conference room lighting utilizes two strong visual elements; an uplighted stepped ceiling coffer and wall sconces. The ceiling coffer provides indirect diffused light to fill the room and create a relaxed atmosphere. Various faces of the steps in the coffer were painted in graded increments of grey (darker toward the bottom) to permit the use of simple fluorescent striplights and to balance brightness in the coffer. The paint values were selected in a full scale mockup (made with foam core) in the lighting designer's laboratory. The low brightness PAR-38 downlights, controlled separately from the uplights, were located in the top of the coffer to provide focused light on the table. Wallwashers at the end of the room provide light for presentations.

An uplighted stepped ceiling coffer and wall sconces dramatically light the conference room.

*Lighting outside the boardroom is
subdued and draws attention to the
brighter boardroom.*

Walker Dental Office

Project: Walker Dental Office
Location: Las Vegas, NV
Interior Designer: Tony Grant, A.S.I.D.
Photographer: Patrick Bartek

It was a very short entry from the front door to the transaction counter—only about 7 feet. It's that short because of the proportion of the suite the dentist had.

The designers tried to open it up as much as possible. To achieve that... notice on the left hand side, there is a pyramid-topped area that is the patient bathroom. They created an inside pyramid roof to the bathroom and lighted above it the same way as they lighted above the reception area, which is just beyond the transaction counter.

There were two advantages to this—symmetry from one side to another, and gave depth needed. Transaction area was already deep enough. Front door to transaction counter was another 7 feet. Beyond that is another 9 or 10 feet for the reception area.

Lighting effect off the pyramid is a nice touch—recessed incandescent downlights were used. Indirect coffers are fluorescent—using two tubes.

The pyramid shape adds architectural interest to the office.

Recessed incandescent downlights provide pleasing lighting.

The pyramid shapes and the associated lighting give the office a nice, symmetrical look.

Burlington Northern

Project:	Burlington Northern, Inc. Corporate Headquarters
Location:	Seattle, WA
Interior Designer:	PHH Neville Lewis
Lighting Designer:	Horton-Lees Lighting Design Inc.
Photographer:	Hedrick-Blessing
Fixture Manufacturers:	Columbia, Linear Lighting, Edison Price, Kurt Versen

The "desert like" atmosphere of the reception area for the executive floor is marked with a large marble sculpture of a wagon wheel which was conceived by the artist in the light of the Arizona sky. The designers tried to recreate the quality of that light through the use of an artificial skylight with high color rendition fluorescent lamps configured into cool and warm zones. The photocell on the roof transmits information to the control systems which appropriately switches illumination between these zones. The shape and design of the skylight reflected the pyramidal sandstone base of the sculpture.

Inside the boardroom the sculptured ceiling and pendant light fixture respond to the strong form of the massive boardroom table below. The dome above is illuminated with fluorescent lamps while incandescent accentlights highlight the table.

The Executive Dining Room is a retreat brightly decorated with antiques and rich woods. Lighting in this room was achieved with incandescent downlights and accentlights to provide the sparkle on the table and an intimate feeling for casual luncheons and meetings.

The connecting stair between the two executive floors floats in a sculptural form and is a backdrop for a hanging plant arrangement. Downlights where strategically placed in patterns to light the stair reception desk, art and seating groups. The lighting throughout the executive floors was controlled by a very sophisticated dimming system with sensors which discreetly raise and lower the light levels as one walks through the spaces.

Incandescent downlights and accent lights work effectively in the executive dining room.

Downlights light the stair, reception desk, and selected pieces of artwork.

In the boardroom, the sculptured ceiling and pendant light fixture correspond to the massive table below.

An artificial skylight adds atmosphere to the reception area.

United Bank of Denver

Project: United Bank of Denver
Corporate Offices
Location: Denver, CO
Interior Designer: Gensler and
Associates/Architects
Lighting Designer: Horton-Lees Lighting
Design Inc.
Photographer: Hedrich-Blessing
Fixture Manufacturers: Columbia Lighting

The building was designed by Philip Johnson. Gensler and Associates were the designers for the bank's corporate headquarters complex. The employees' cafeteria consists of several "theme" rooms, one of which is shown. The intent here was to create the feeling of being outdoors on a bright sunny day, sitting in the shade of a large umbrella.

The effect was achieved by intense incandescent downlights laid out so as to limit their beam spread to the umbrellas and to minimize their visibility from the passageways.

The design intent in the boardroom was to have the lighting system echo the shape of the table. The lightsource is a large luminous panel formed into a horseshoe similar to the table. High color rendition of tri-phospor fluorescent lamps enhances the appearance of occupants and materials used in the space. The fixture is equipped with a light-colored aluminum eggcrate louver and is dimmer-controlled.

To create an outdoors feeling, intense incandescent downlights hit the umbrellas, creating the feel of a sunny day.

The lighting echos the shape of the boardroom table.

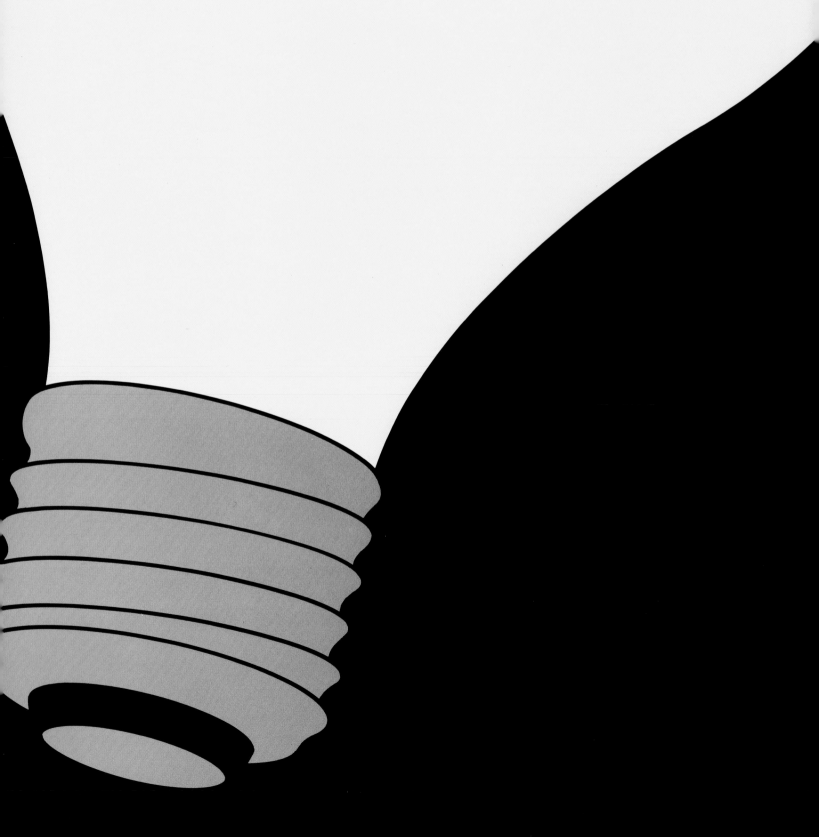

CHAPTER 5
PUBLIC AREAS

The main lobbies of many buildings and corporations must be well lighted to properly illuminate directories, exits and entrances, elevator banks, stairwells and various forms of signage. In addition, these areas must be properly illuminated so pedestrians can clearly see, so that the traffic flow remains smooth and steady.

These areas also serve as information spots—they are perfect spaces to showcase exhibits, post notices and display artwork.

This chapter contains projects that effectively light public areas—the imaginative lobby at 388 Market Street in San Francisco, California; the banking platform at Commerzbank in New York City; and the breathtaking lobby at Pacific Plaza in California.

ANZ Bank

Project: ANZ Bank
Location: Melbourne, Australia
Lighting Designer: Horton-Lees Lighting Design, Inc.

The project is located in a multipurpose development containing in addition to the bank hall, two office towers, a hotel, large indoor/outdoor atrium and shopping arcades.

Featured is a two-story high entrance hall with the circular information desk as its focal point. Incandescent downlights and wallwashers are used to emphasize the form of the total space and the shape of the curving upper level balcony. Large ceiling recesses house directional accentlights which illuminate the main floor circulation areas without scalloping the vertical surfaces of the opening.

Two escalators traverse the space and connect the upper level of the entrance hall with the lowest (third) level of banking hall. The grid of low brightness downlights is augmented by a row of wallwashers illuminating the 60-foot high back wall.

A grid of downlights and wallwashers illuminate the 60-foot high back wall.

Incandescent downlights and wallwashers are used to emphasize the two-story entrance hall.

Shorebreeze

Project:	Shorebreeze Lobby
Location:	Redwood, CA
Lighting Designer:	Luminae, Inc., James R. Benya, Principal-In-Charge
Interior Architect:	Ehrlich Rominger, Laura Seccomb, Project Designer
Photographer:	Douglas Salin
Fiber Sculpture:	Judith Content

The aim of the developer was to improve the image of the 120,000 square foot, multi-tenant building through the use of fine finishes in the public spaces —the two-story atrium features beautiful marble and tapestry.

General lighting is provided by a combination of light sources. Each delicate Aurora pendant fixture contains three halogen lamps, with four pendant fixtures suspended within each coffer. Blue cathode cove lighting surrounds the perimeter of each coffer. After-hours security lighting is provided by metal halide lamps recessed into the coffer ceiling. The lower ceilings that run along the perimeter of the atrium are equipped with recessed MR-16 accent light and wallwasher fixtures, and blue neon-uplit coffers.

The atrium's eye-catching attraction, the 12-foot high tapestry, is illuminated by four mini-ellipse fixtures concealed and used as framing projectors. Elevator lobbies are highlighted with two matching Italian blue glass sconces and ceiling coffers.

At the atrium's two levels, the coffers are blue uplit; on other floors, the coffers are warm white uplit in cold cathode. Signage on all floors and elevator lobby carpet inlays are accented by recessed MR-16 adjustable fixtures.

General lighting is provided by a combination of light sources: pendant fixtures, cove lighting, accent lights, to name only a few.

©1987 Douglas Salin

Commerzbank

Project: Commerzbank
Location: New York, NY
Design Firm: CHS Planning and Design

CHS Planning and Design Corporation used cathode ray lighting to dramatically articulate space in Commerzbank's office in the Wall Street area of New York City. The lighting outlines the edges of the walls, curving around columns and creating freeform patterns in the ceiling.

Cathode ray lighting creates a continuous stream of light, something no other light source is able to do. Like neon, cathode ray lighting tubes are pressurized, containing a potentially dangerous gas. As a result, the light must be handled by someone specifically trained in its installation.

Cathode ray lighting is placed in a cove to increase its reflective qualities. Generally, its use is restricted to high-ceilinged spaces because the lighting emits an unusual amount of heat.

At Commerzbank, CHS maximized the design potential of cathode ray lighting by bending the light around workstations and entranceways, providing necessary illumination while enhancing the design elements of highly reflective surfaces and curved furnishings. The use of cathode ray lighting added dimension and style to this corporate setting.

Cold cathode patterns are pleasantly distorted in dark tinted plastic shields.

Cold cathode cove forms a band around the perimeter.

388 Market

Project:	388 Market Street
Location:	San Francisco, CA
Interior Designer:	Skidmore, Owings & Merrill
Lighting Designer:	Horton-Lees Lighting Design Inc.
Photographer:	Jane Lidz
Fixture Manufacturers:	Shaper Lighting: Pendants and sconces; Edison Price, Inc.: Downlights; Brumfield Electric Sign Co.: Cold cathode

This speculative office building is located in the financial district of San Francisco. This building conformed to the San Francisco downtown plan which required the building to be partially residential. It is of new construction and the scope of work included the main floor public spaces, the second floor elevator lobbies (which crossed the main lobby via two skybridges) and the residential reception and elevator lobbies.

In the two-story main lobby, the floor and wall finishes are polished granite and marble, with polished stainless steel metal trim. The shallow vaulted ceiling is panelized and has a stark white finish. The goal to provide a luminous, inviting environment was complicated by the largely polished, dark surfaces. The designers wanted to accentuate the architectural details as much as possible and selected an indirect scheme for this area. They developed a pendant uplight assembly which houses four metal halide uplights, each having asymmetric forward throw reflectors. The uplights are aimed along the axis of the ceiling, providing a very uniform ceiling brightness as well as lighting levels of approximately 15-20 footcandles at pedestrian height. To provide some perimeter brightness, custom luminous wall sconces are located along the walls at the ground floor level and at the second floor level to mark the entries into the corridors and elevator lobby. The sconces are fabricated of white acrylic which was sandblasted and sealed on the exterior surface. The metal trim is polished stainless steel. Because of the size and shallow depth of the sconce, compact fluorescent lamps were used which uniformly illuminated the diffuser.

The fixture also has a top diffuser to shield the sconces as you can see directly into the sconces from the sky-bridges above.

The 16-foot tall elevator lobby is also indirectly illuminated with a narrow light "slot." The goal was to illuminate this slot and not be able to see the corners of the cove above. This was achieved by doing comprehensive sight line studies and using fluorescent lamps hidden within the architectural cove.

At the two ends of the elevator lobby are similar circular coves which are illuminated with cold cathode. The upper cavity of the cove is domed to provide a uniformly illuminated cavity and to eliminate views of any seams or corners.

Ambient illumination levels are approximately 15-20 footcandles. The wall sconces are continued in this space from the main lobby.

A narrow light slot indirectly illuminates the elevator lobby.

This office/residential building is located in downtown San Francisco.

A number of light sources work together to provide general illumination.

Pacific Plaza

Project:	Pacific Plaza
Location:	Pleasant Hill, CA
Architect:	Robinson, Mills, and Williams
Lighting Designer:	Horton-Lees Lighting Design Inc.
Fixture Manufacturers:	Edison Price, Inc.: Accentlights and downlights; Lightolier: Wallwashers; Shaper Lighting: Custom fixtures; Neoray: Windows
Photographer:	Paul Peck

This speculative office building is located in Pleasant Hill, Caliornia, about twenty-five miles from San Francisco.

The main floor lobby is composed of four attached rectangles which bring a person from the two exterior entries to the main lobby and from there to the elevator lobby. The primary entry is open through two floors and is visible across a long pedestrian plaza from the outside.

The designers were originally asked to analyze an artificial skylight scheme and felt that although they could create a comfortable level of light, it would be such a large skylight that the space as a whole would become very bland. Instead, they created pendant mounted fixtures with details similar to the screens around the perimeter. They are supported from the center so that from below they appear to be floating. The source is fluorescent. The lamps are mounted on special strip with chamfered edges to eliminate any shadows on the glass diffusers. The fixture is steel construction with clear glass that is airbrushed white on the inside and sand-blasted on the outside, the lamps are tri-phosphor type. This project meets California Title 24 energy standards.

The designers created pendant mounted fixtures that appear to be floating; a beautiful effect.

A view of the main lobby.

Bank of Dallas

Project: Republic Bank of Dallas NA

Location: London, England

Lighting Designer: Sally Storey, Lighting Design Limited, London, England

Architect and Interior Designer: David Leon & Partners

Photographer: Bob Belton

In the main banking hall, tungsten halogen power floor luminaires are set into the cornice at the top of each painted column and provide indirect lighting. Low-voltage tungsten halogen spotlights are used around the edge of the hall. They highlight the entrance and artwork, and provide additional lighting in the seating area.

Simply furnished with rich materials, there is no glare in the floor marble from overhead sources because of the indirect lighting.

The architectural elements are allowed to dominate the space because the lighting is kept hidden from direct view —recessed in the top of the columns. Small spotlights are used to highlight artwork and to cover the seating area.

California First Bank

Project: California First Bank
Location: San Francisco, CA
Architect: Skidmore, Owings & Merrill
Lighting Designer: Horton-Lees Lighting Design Inc.
Photographer: Jaime Ardiles-Arce
Fixture Manufacturers: Columbia: Building Standard Lighting; Lightolier: General office floor light tracks; Edison Price, Inc.: Downlights for banking hall and mezzanine

The banking hall complex includes the main banking hall (shown), the building lobby and elevator lobbies. The design of the large hall (60 ft. x 60 ft. x 40 ft. high) is a modernist counterpoint to the ornate space of the Bank of California across the street; the main motive could be called "circles in the square." The center of the space is occupied by a circular tellers' counter; a second circle above it is formed by a suspended task fixture—a custom structure containing standard incandescent downlights.

Inside the tellers' counter is a smaller circle of backcounters illuminated by built-in cold cathode strips formed to suit the backcounter curvature.

In the center of the inner circle stands a cylindrical shaft of the hydraulic elevator leading to the banks' vault in the basement. The top of the shaft contains a group of color corrected mercury vapor uplights which provide ambient illumination for the space.

The linear slots in the ceiling contain adjustable downlights which provide general illumination for the circulation areas as well as more intensive task lighting for the officers' platform on the ground floor and the miscellaneous office spaces on the mezzanine. These downlights are mounted to the catwalks in the mechanical space above the ceiling.

A grid of small recessed incandescent downlights illuminates the workspace under the mezzanine.

The linear slots in the ceiling contain adjustable downlights for general illumination.

Suspended task fixtures illuminate the tellers' area.

A grid of small recessed incandescent downlights illuminates the area.

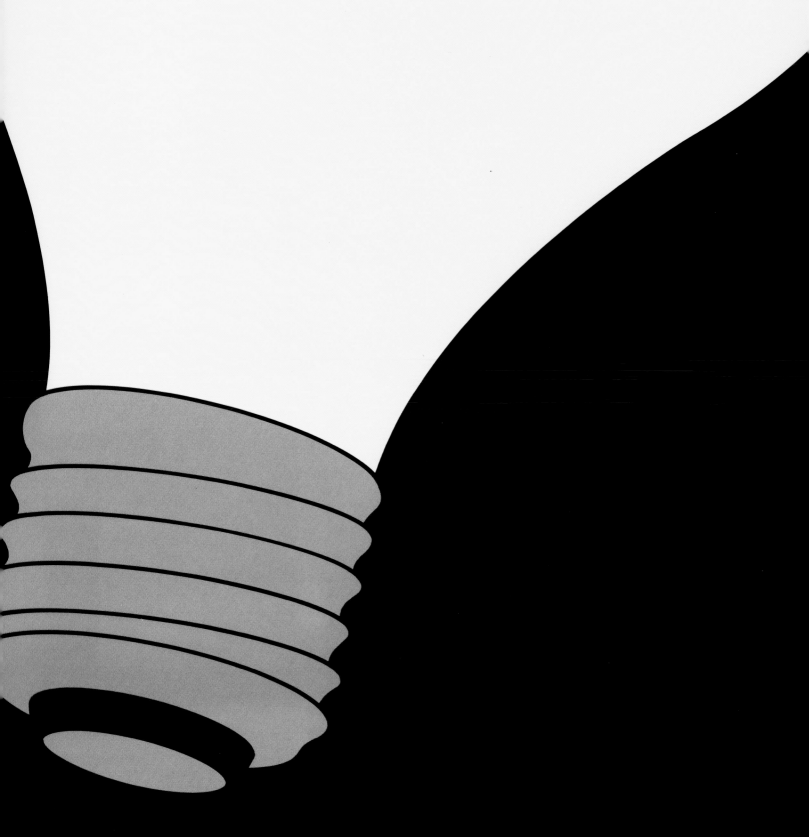

CHAPTER 6
FUTURE TRENDS

As today's office incorporates newer and more technologically advanced equipment, lighting systems must change and adapt to meet these needs. For example, glare on computer terminals was not a lighting consideration 15 years ago. Today, it is a major concern that must be addressed when a designer works on any project involving computer equipment.

This chapter contains projects that utilize some of the latest lighting technology available (unusual use of fabric panels and lighting equipment is illustrated in the Tensilight System Project). In addition, this chapter features one of the hottest trends in business today—the home office (portable and artistic lighting is showcased in the Weiss Home Office Project). As cottage industry continues to bloom, and as commercial rents continue to rise, the need for a well lighted and functional home office will continue to be a challenge for many designers.

Tully Weiss Home Office

Project: Tully Weiss Lighting
                     Design Home Office
Location: Dallas, TX
Lighting Designer: Tully and Kalynn Weiss
Interior Designer: Tully and Kalynn Weiss
Photographer: Ira Montgomery

This office is used to show clients examples of side backlighting, key, fill, filter, color, fresnel, ellipsoidal reflector spotlight, different types of low voltage and line voltage, tungsten halogen, incandescent lamps and other various ways of using artificial light.

The furniture in the reception/living room was changed every three to four weeks to show and experiment with different lighting effects and compositions.

Portable lighting trusses with various miniature equipment was used to high-light the room and furniture. Floor floodlight scoops were used to fill the room with bright, colorful fill light.

Tully Weiss uses theatrical technology to give homeowners the ability to live in a space filled with vivid color and excitement. Different types of lighting can be combined so that it creates a stimulating, visual atmosphere. Tully Weiss feels that the best type of lighting offers contrast.

Ceiling mounted ellipsoidals light the seating area in the bedroom.

In this home/office portable lighting fixtures were used exclusively throughout.

Parachute material provides a dramatic backdrop for illumination.

Supercomputer Center

Project: John von Neumann Supercomputer Center
Location: Princeton, NJ
Lighting Designer: CUH2A, Princeton, NJ
Interior Designer: CUH2A, Princeton, NJ
Photographer: Robert Faulkner

The John von Neumann Supercomputer Center of The Consortium for Scientific Computing is used by researchers from member organizations all over the world. The needs of the highly educated user group, combined with the sophisticated state-of-the-art equipment being used at the supercomputer center, dictated some dramatic changes to this one-time spec office space. The lighting system in particular needed to be completely redesigned.

Efficient glare-free lighting was the primary objective in the work-intensive research and computer areas. Throughout the corridors, break areas and lobby, a comfortable and pleasant environment that provides visual relief was the objective. The facility's high visibility to the international research community required the lighting system to reinforce the building's aesthetics, helping to create interesting and appealing spaces.

The typical research office is a 10' x 15' enclosed space located on the building's perimeter. Each office is equipped with computer terminals, printers and other high tech peripherals. A task-ambient lighting system illuminates this space. Pendant-mounted aluminum uplights with T-8 fluorescent lamps and specially designed low-brightness lenses provide low levels of ambient light. Adjustable task lights equipped with twin tube fluorescent lamps punch up illumination at desk areas. Extensive computer modeling was conducted to predict the complete luminous environment, and a full-scale mock-up was evaluated as part of the design process.

The computer training center is an open plan in the center of the building that provides visual relief for the researchers and a high profile area for visitors touring the facility. Indirect tubes, similar to those in the research offices, are arranged in geometric patterns within ceiling coffers to illuminate this space. A higher level of ambient illumination is provided by these fixtures because there are no task lights. Computer modeling was also used to develop the design.

The computer room is also centrally located and highly visible. Although it is an enclosed and secure area, window walls along the perimeter allow a complete view of one of the world's most powerful computers. Recessed parabolic troffers provide high levels of illumination both to display the equipment and light a variety of visual tasks.

The corridors circulate throughout the building, separating the perimeter research offices and the central computer and training areas. The long hallways are broken up by lounge seating and break areas that view the computer and training areas and give researchers an interaction space. Continuous fluorescent indirect tubes light the peaked corridors, reinforcing the lighting vocabulary and emphasizing the architecture. Incandescent wall-washers and accent lights add warmth and a relaxing residential quality to the seating and break areas.

The main lobby is the primary entrance and exit to the building. Both researchers and visitors access the facility through this space. Indirect cove lights and wall sconces provide overall ambient illumination and establish the indirect lighting approach. Incandescent wallwashers add warmth to the seating areas and accent lights highlight the glass-etched Consortium logo.

Continuous fluorescent indirect tube lights light the corridors.

In the computer room, recessed parabolic troffers provide high levels of illumination.

Indirect tubes are arranged in geometric patterns within ceiling coffers to illuminate the computer training center.

Indirect cove lights and wall sconces provide overall ambient light in the main lobby.

IMNET

Project:	IMNET
Location:	Princeton, NJ
Lighting Designer:	CUH2A, Princeton, NJ
Interior Designer:	CUH2A, Princeton, NJ
Photographer:	Wolfgang Hoyt, ESTO Photographics

International MarketNet, a joint venture of IBM and Merrill Lynch turned a two-story 72,000-square-foot standard spec office building into high tech office space that supports large-scale computer use, corporate offices and electronics fabrication.

IMNET made the decision to relocate to the existing Princeton Forrestal Center office building in April 1985. At that time, the company was occupying offices in Merrill Lynch's New York headquarters. Renovation of the new facility began in September, and 167 employees were moved to Princeton following six separate construction phases.

IMNET wanted to save as much of the existing building infrastructure as possible in order to cut construction costs. Usable existing interior walls and mechanical ducts were retained, as were the core area support services--elevator, restrooms and mechanical systems—in the center of the building. The building was stripped of existing carpet, ceiling, column covers, blinds and finishes.

Building around existing core services, the new IMNET facility is laid out like a figure eight, with the lobby and cafeteria on opposite ends of the first floor. Inside the core spaces on the first floor are the remaining support areas, a computer control center, a communications room, a peripherals room, and a 16,000-square-foot maintenance and distribution center where IBM PCs are modified and repaired. In the core area on the second floor is a 6,000-square-foot computer room.

Outside of the figure eight on both floors is 23,000 square feet of open office area with 8' x 9' modular work-stations. Enclosed 10' x 12' managers' offices, six 14' x 15' executive offices, and multiple conference rooms are located along the perimeter windows. Informal meeting areas are located in the corners of the building by large windows.

IMNET wanted a quiet office. The sound-absorbing ceiling tiles in the open office areas have an STC rating in the upper 30s and an NRC coefficient of 0.5-0.6. Fabric-covered partitions between workstations have an STC rating of 14 and an NRC coefficient of 0.8. The noisy telephone hotline room uses ceiling tiles with an STC rating in the low 30s and an NRC coefficient of 0.9-1.0.

The large number of CRT screens in the facility also demanded special lighting. Parabolic 1½-inch cubes are used throughout the office and computer areas. CUH2A project electrical engineer Robert Marshall says the cube lighting helps to even out the light and dark spots of the ceiling reflected on the CRT screens. Corridors are lit with energy-efficient fluorescent lamps.

A combination of light levels was designed for the reception area at IMNET—both the lower light level needed for VDTs as well as incandescent downlights and wallwashers to brighten the overall environment.

The indirect fluorescent lighting provides illumination without throwing glare onto the VDT screens.

The lighting system throughout the office is a combination of extruded indirect fluorescent lighting, recessed small parabolic fluorescents, and task lighting affixed to the undersides of furniture cabinets and shelves.

Lane Home Office

Project: Lane Lighting Design Home Office
Location: Seattle, WA
Lighting Designer: Lori and Michael Lane, Lane Lighting Design

Last year Michael and Lori Lane started Lane Lighting Design, Architectural Lighting Consultants. Because of their desire to work at home, they purchased a house in which 500 square feet could be an office. The home had three bedrooms upstairs which were renovated into an office/studio.

Being on a tight budget, Lori and Michael tackled all the work themselves. Also, most of the furniture was designed and constructed by them. The space is tailor made to fit workstation needs.

After the plans were drawn—demolition began. All non-structural room partitions were removed to open up the space. The central part of the ceiling was removed in order to install the light fixtures and also to place insulation.

Since they were dealing with angled ceilings—custom details were necessary. Michael designed and built the bookcase located in the central conference room.

Since their profession is lighting design consulting, they wanted to install innovative lamp sources and fixtures that their clients could view. The recessed downlights use a compact fluorescent lamp source. They consume only 22 watts of electricity and provide the same illumination level as a 100-watt incandescent lamp. The bookcase is illuminated with a custom built fluorescent valance in which light grazes the catalogs. The pendant fixture over the conference table uses a silver bowl incandescent lamp for a soft glow of light.

For a design firm, storage of drawings and product samples can take up lots of space. The solution was to use the eve space, accessible from the end offices. These spaces proved very useful for wiring access when installing the telephone and computer system.

The outdoor deck provides a nice getaway to the office/studio. It is a great place to relax and brainstorm plus enjoy listening to the birds sing, one benefit of a home office.

This view is facing east looking into Lori's studio space. The recessed downlights use the compact fluorescent lamp.

This view shows the computer equipment set on a desk made from a hollow core door and painted sawhorses.

This view is from the top of the stairs looking into the Conference Room. Notice the light grazing the catalogs, an integral part of the custom built bookcase.

This view faces west showing the copier in the foreground and the deck beyond.

Lightolier Case Study

Project: Major Utilities Company
Location: Denver, CO
Lighting Designer: Lightolier
Architect: Nora Dimatron
Electrical: Kirk Davis, Cator-Ruma & Associates
Products: Lightolier Lytespread/6 Architectural downlights

Cost

Previous indirect fluorescent systems had cost the owner $35-$40 per linear foot. Coupled with the fact that these systems lacked flexibility, the owner was extremely concerned about the cost of the system. Through work with the local factory agent, they were able to assure the owner that the system could be purchased for $20-$22 per foot.

Light Levels

Light levels for VDT tasks and pencil tasks are in direct contradiction with each other. Adequate levels for pencil tasks, misdirected, can cause literal

headaches for VDT users. As a consequence, the owner chose to keep the overall light levels between 30-40 foot-candles, with task lights integral to the workstations providing increased levels for pencil tasks. The indirect system installed provided the overall ambient level desired, at a watt density of .9 watts/sq. ft.

Glare

As previously stated, poor lighting can be a large problem in VDT viewing. Glare, or more properly the lack of it, was a major design concern of the team. In theory, an indirect system, provided it

The lighting design of the space resulted from a mutual concern of all team members for a solution which would enhance visibility of both VDT and pencil tasks at each workstation. It was the consensus of the design team that indirect lighting could be successfully employed to provide that visibility. However, it was equally agreed that totally direct systems were psychologically uncomfortable. Lacking a direct component, totally indirect systems were shadowless and therefore dreary. Since most interaction between individual occupants occur either in conference rooms, private offices or in the aisles on the way to and from the coffee area, it was agreed to provide a direct/indirect fluorescent. Private offices were provided with low brightness fluorescent troffers. The aisle ways were highlighted with incandescent downlights.

The criteria used to evaluate the overall indirect systems were:
- a. Flexibility
- b. Cost
- c. Light Levels
- d. Glare
- e. Occupant Orientation

Flexibility

The owner utilizes a modular furniture system in almost all of their buildings. As such, spaces frequently change. These changes include floor to ceiling walls being added to previously open office areas. Static, non-flexible indirect systems were ruled out. Consequently, a modular, plug-in system was selected. The owner was provided with a stock of spare components as part of the bid to allow for this future flexibility.

gives even ceiling brightness, should reduce glare on the VDT screens. Computer studies had confirmed that the system proposed would provide even ceiling brightness. To satisfy the architect and owner, however, a nearby site was located where the indirect system had been previously installed. The design team visited the site, recorded foot-candle levels, and observed the ceiling brightness ratios and VDT screen visibility. All aspects of the system observed, met or exceeded the design team's expectations.

Occupant Orientation

Ironically, one problem with open office landscape systems is their vast open areas. Occupants often have a reduced sense of belongings. When regimented patterns of troffers or long, continuous expanses of indirect fluorescent luminaires are superimposed upon the open area, the problem is exacerbated. In an effort to define smaller "open areas" the lighting system and the workstations beneath, were rotated 45° in each quadrant of the floor plate. This created four separate and distinct open areas.

Cove lighting provides a dramatic enhancement of the curved walls.

Wallwashers highlight art on side walls.

Fluorescent lighting provides the major source of indirect illumination.

Task lighting is available under shelves and furniture cabinets at individual workstations.

Tensilight System

Project:	Tensil Lighting System for Sunar Hauserman
Location:	Cleveland, OH
Lighting Designer:	Peter Barna, 989 • 89— New York, NY
Architect:	Nicholas Goldsmith, FTL Associates— New York, NY
Photographer:	Durston Saylor
Lighting Manufacturers:	Sunar Hauserman

Shrinking personal space has moved workers far from the light of our natural world. This project structures space by fusing light and form at an environmental scale. By doing so, broader visual and sensory needs are addressed.

An open office furniture system was selected as an exemplary structure for the concept.

A family of three lighting/tensile elements fit modularly into the system:

• *Canopy Structure:* A square canopy provides roof enclosures and spatial definition for private offices and conference rooms. System-mounted uplights provide sufficient ambient light for office and conference functions. Pultruded fiberglass wands stress a tent-like form. This extremely lightweight canopy can be supported from four corners or cantilevered from two corners as show in the prototype. Uplights allow for flexibile field relamping for incandescent, quartz halogen or miniature fluorescent sources.

• *Wing Structure:* The wing structure starts at the top of the system panels and fans out as it moves upward to provide a partial sense of enclosure for workstations arranged from a central spine. Structural arms inside the wing enclosure extend from the system's tubular structure and provide tension at the top edge giving the wing its characteristic shape. Miniature fluorescent sources inside the wing provide a diffuse, ambient light.

• *Light Panels:* The light panels can replace any of the existing opaque fabric wall panels. Calendered nylon is stressed into an aluminum frame by thin vertical ribs 1' on center. Miniature, 5-watt fluorescent lamps behind each rib provide the appearance of a column of light within each panel. Because the light is diffused over the large surface area of the panels, discomfort glare is prevented. At the same time, lighting levels on the work surfaces are appropriate. The modular system allows for placement of luminous panels to the sides of work surfaces preventing veiling reflections and CRT screen glare.

Scores of fabrics were tested for their optical characteristics and ability to be stressed. The selected nylon is flame treated, machine washable, durable, lightweight and provides high light transmission and diffusion. A variety of lighting sources and locations were tested with single-ended, fluorescent sources providing the best overall appearance, the least energy consumption and the longest life.

Inherent in the Tensil Lighting Environment is the flexibility of light source location. This is important because light source location is the most critical element in prevent CRT glare. The three system components are modularly sized as established by Douglas Ball when he designed the Race Furniture System.

Marketing Studies presently underway indicate that component costs will be very competitive with existing lighting systems.

By combining light and form into single components, the Tensil Lighting Environment increases the functional efficiency of design and brings a strong aesthetic identity to the workplace. It is an identity with which the organization and the individual workers can relate. The range of scale of the unique components sets it apart from other systems presently marketed. The light panel alludes to a luminous window and provides a personal scale. The scale and form of the wing and canopy interject a dynamic quality to the rhythm of the grid. They allude to shelter and sky; and, because they are luminous, contribute to, a sense of well-being and comfort.

In summary, the Tensil Lighting Environment provides a functionally efficient, aesthetically pleasing and unique sense of place which satisfied broad visual and sensory needs.

The curved forms of the lighting wing and the light panels are computer generated.

The three elements of the Tensilight Lighting system: the canopy, the light wing, and the luminous panel. Attached reflector uplights the canopy.

The moveable luminous panels integrated into the wall system provide task lighting.

CHAPTER 7
NEW PRODUCTS

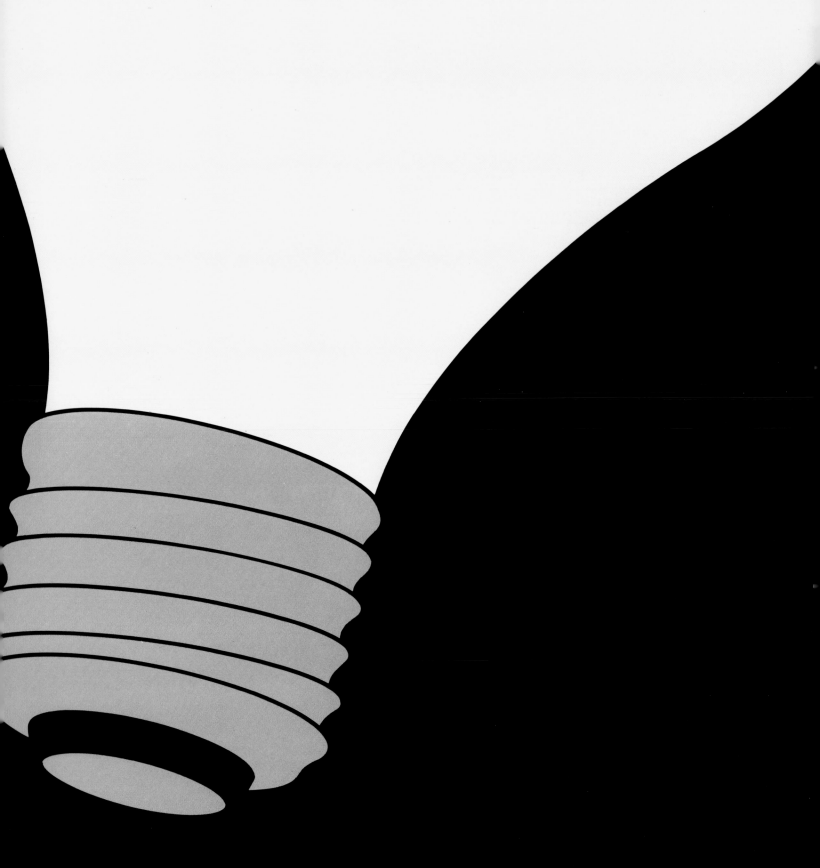

The new products featured in this chapter are not intended to be a comprehensive collection of all that is new in the lighting marketplace. They merely represent a selected sampling of various new products currently available to the designer and consumer.

Many of the products shown reflect the philosophy of current design trends—to improve and refine existing lighting equipment design, to enhance efficiency and to promote cost and energy savings.

Another trend is flexibility. Many fixtures must perform a number of lighting functions and still retain an esthetic sense. No longer will a bare bulb suspended from the ceiling suffice.

Finally, today's lighting products must be able to satisfy specific needs. The increase of computerization in the workplace has brought with it many lighting requirements and restrictions that must be met in order that employees can function at peak productivity.

Aamsco
Manufacturing Inc.

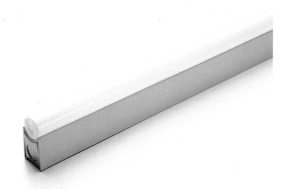

Alinea linear incandescent luminaire radiates a soft, warm light at 2,800 Kelvin that is flattering to fleshtones. It is controlled by a standard incandescent dimmer, needs no transformer or ballast, and works off standard line voltage. Fixtures are available in polished black enamel, white enamel, satin finished silver, antique gold, fire engine red, electric blue, sage green and sun yellow.

Alinea has a channel housing that is contoured to surround the bulb. The fixture and bulb come in three lengths: 11.8 inches (300 mm), 19.68 inches (500 mm), and 39.37 inches (1000 mm). End knockouts allow a continuous line of light to be created by mounting Alineas end-to-end. Modular design makes bulb replacement fast and easy.

Alkco

The Recessed Track TM integrates the fixture placement and aiming flexibility of track lighting into a system which is hidden from view. Low voltage MR-16 lightmodules can be aimed 0 to 40 degrees, rotated 380 degrees, and placed anywhere along the length of the Recessed Track. The track and lightmodules are concealed from view, recessed above a cleanly detailed linear trim. All of the light passes through the narrow, 2-inch linear aperture, regardless of the lamp aiming angle. Different lamp wattages and beamspreads allow for a variety of effects. The track can be run in continuous lengths and patterns.

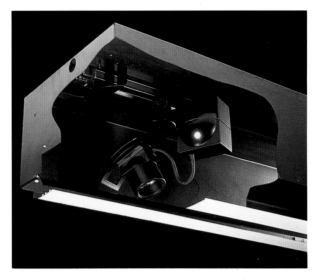

In the Recessed Track system, dual circuits allow separate switching/dimming at individual lightmodules located in the same housing. The aperture cone reduces direct glare and minimizes spill light around the periphery of the beam pattern. The hood assembly is easily removed for relamping. An integral 12-volt transformer properly operates the entire range of MR-16 lamps from 20 to 50 watts. Sturdy die-cast metal construction promotes optimal thermal dissipation.

Varilux TM three-level task light integrates two lamps (13-watt PL flurorescent) into a single luminaire. A specially-designed reflector and lens direct light laterally from the sides and across the workstation to minimize glare, veiling reflections and shadows. A recessed retainer hides chords. The fixture is easily mounted to the under-side of metal shelves without tools, using a quick-install mounting bracket, with double-faced adhesive strips.

Wallscapes TM integrates high-efficiency, superior-color fluorescents into a compact wall-mounted lighting system. Different optical packages can be specified to provide flexibility for direct, indirect, or combination direct/indirect lighting applications. Wallscapes are available in 1-, 2-, 3-, 4-, and 5-foot lengths. One-foot units can be used as sconces to provide evenly distributed accent lighting without hotspots on the wall. Wallscapes can be mounted end-to-end for continuous run of uninterrupted illumination in applications which require indirect or perimeter lighting.

Alkco

The Lumenizers TM fixture group integrates the HQI lamp into a versatile, compact, architecturally styled fixture. Three optical packages are available: Lumenspot TM produces a soft-edges, round beam for lighting objects and small displays; Lumenflood TM produces a broad rectangular pattern for large display and area illumination; and Lumenwash TM has an asymmetrical distribution for even illumination of vertical surfaces. All Lumenizers are available in surface, semi-recessed, pendant and portable mounting configurations for use indoors. Surface units can also be specified with a UL wet location listing for use outdoors.

Artemide, Inc.

The Platone ceiling lamp, designed by Ettore Sottsass, is 35cm wide with a 46cm overhang. The painted white metal lamp uses a maximum of four 100-watt clear (E 27) bulbs.

The Tiara wall lamp, designed by Gianfranco Frattini, is 25cm wide and 11cm high with a 22.5cm overhang. The white molded and glazed glass lamp uses a maximum 300-watt halogen (R7s/15) bulb.

The Aton Modular Lighting System, designed by Ernest Gismondi, is available in white, black, Chinese red and anodized aluminum. The Aton Modules are made of extruded aluminum with a lacquered finish, and are available in a variety of sizes with fluorescent, halogen or incandescent light sources. A variety of connectors allow maximum layout flexibility and easy installation. Accessories available to equip the system include signage, electrical outlets, and speakers.

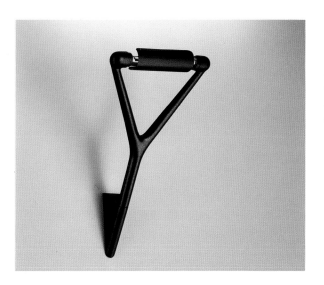

The Lola lamp, designed by Paolo Rizzato and Alberto Meda, uses one Q300 T3/CL. Lola is made with new high technology materials, such as carbon fibers, flexible self-skinned polyurethane, and thermoplastic polyester.

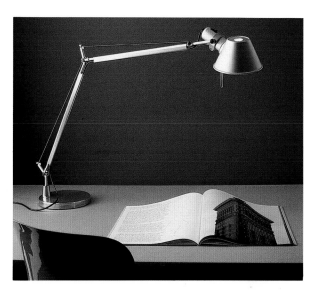

The Tolomeo, designed by Michele De Lucchi and Giancarlo Fassina, is a fully adjustable task lamp in polished aluminum with a base of black die-cast metal alloy. Tolomeo uses a maximum of one 100-watt white (100A) bulb or one 50-watt quartz halogen bulb. The halogen model is available with a standard reflector or a special anti-glare, parabolic, "dark light" reflector for task lighting at computer terminals.

Artemide, Inc.

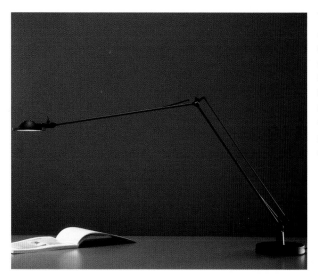

The Berenice table lamp, designed by Paolo Rizatto and Alberto Meda, uses one 35-watt, 12-volt halogen bulb. The lamp's metallic parts in pressure casting of aluminum are available in black or silver finishes. Adjustable hinges are reinforced nylon. The parabolic reflector is available in aluminum or green pressed glass. Berenice is also available in clamp-on, wall, and floor models.

Atelier International Lighting

Moni is a ceiling fixture designed by Italian architect, Achille Castiglioni, to resemble a shining star in the night sky. Lighting from a reflector, either 13.5 or 17.5 inches in diameter, is emitted through evenly spaced metal rods to create the starburst effect. The round center section of convex glass surrounded by a larger ring of opalescent glass directs additional light downward. In addition to residential applications, Moni is suitable for conference rooms, corridors and building lobbies. Moni's light source is a 100-watt incandescent bulb.

The fan-shaped Butterfly Wall Sconce, designed by Italian architects Afra and Tobia Scarpa for Flos of Italy, radiates variable-intensity light through both frosted glass and woven fabric diffusers. Lighting effects are varied by opening or closing the white fabric diffuser or by repositioning the lamp on its central axis. The cast aluminum base is available in black, pearl white or mauve; the diffuser is composed of woven and pleated, flame-resistant white fabric. Two elliptical frosted glass panes act as bulb guards for the lamp's 150-watt halogen lamp.

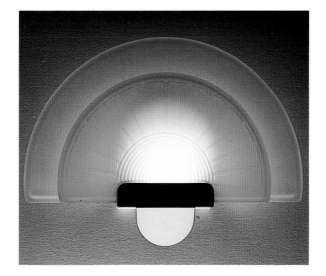

The Diva wall sconce, designed by Italian architect Ezio Didone, has two semi-circular glass diffusers attached to a cast aluminum body. The 9 3/8-inch diameter front panel of frosted and patterned glass is available in white or rose colors, with a white, textured rear panel 12 1/2-inches in diameter. The textured white enamel backplate reflects light from a 100-watt incandescent bulb to the rear, while soft diffused light is produced through the front and around the two frosted glass panels.

The Bodine Company

The HID 1600 Emergency Lighting System (patent pending) allows high-intensity discharge (HID) lamps to be used for both normal and emergency illumination, and eliminates the need for obtrusive auxiliary or quartz restrike lighting. Power to lamps is present continuously with this system, so emergency operation occurs without any transfer time. When AC power fails, the DC power supply continues to provide power to independent remote inverter ballasts. If one fixture fails, the others continue to operate at about 50 percent of rated light output, provide a high quantity of illumination familiar light sources, and help reduce occupant panic.

Brueton

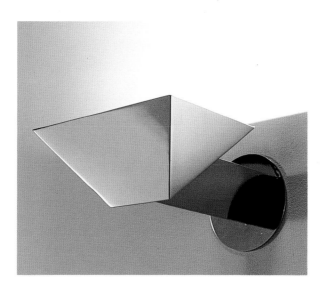

The Delta wall sconce, designed by Alex Forsyth, has a triangular-shaped housing that is 12 inches wide, 10 inches deep, and 5 inches high. The housing and support arm are welded to a circular wall mounting. Light is produced by a 300-watt pure white halogen bulb that illuminates areas above and behind the lamp.

Brueton

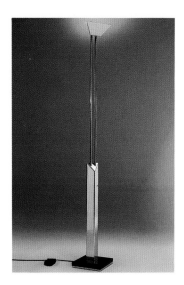

The Delta torchiere, designed by Alex Forsyth, has a stable 9-inch square marble or granite base that supports a triangular-shaped stainless steel pedestal and tubular supports. The supports are welded to a triangular-shaped housing. Lamp pedestal and housing are available in polished or satin stainless steel; the base is offered in 18 stone selections. The torchiere is 74 inches high. The housing is 14 inches wide, 14 inches deep, and 5 inches high. Light is produced by a long-life 300-watt halogen bulb that illuminates areas above and behind the lamp.

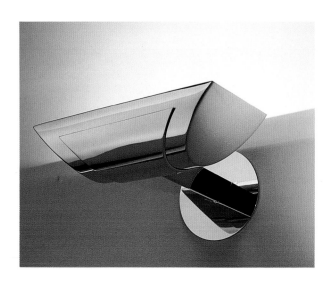

The Theta wall sconce, designed by Alex Forsyth, has a half cylindrical, trough-like shell housing with angled sides that consists of two identical stainless steel sections cut apart and welded together to form a three-dimensional, bas-relief effect. The housing is connected to a stainless steel arm support and welded to a circular wall mounting. The sconce is 12 inches wide, 12 inches deep and 5 inches high. It is available in polished or satin-finish stainless steel. Light is produced by a long-life 300-watt halogen bulb that illuminates areas above and around the sconce.

The Zeta wall sconce, designed by Alex Forsyth, appears as though it is part of the wall to which it's attached. The 4 1/2-inch diameter cylinder, truncated at 45 degrees on both sides, is attached to the wall by a hidden bracket inside the housing. Zeta, available in polished or satin-finish stainless steel, is 10 inches high and 9 3/4 inches deep. The 150-watt long-life halogen bulb provides pure white, direct, upward illumination.

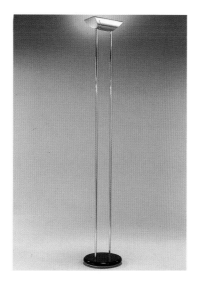

The Theta torchiere floor lamp, designed by Scottish industrial designer, Alex Forsyth, has a stable, 12-inch round, bullnose-edged marble or granite base connected to the lamp housing by two slender 5/8-inch round, stainless steel upright supports. The housing is a half-cylindrical trough-like stainless steel shell with angled sides. Two identical sections are welded together to form the three-dimensional, bas-relief effect of the housing. The upright supports and housing are available in polished or satin-finish stainless steel; the base is offered in a choice of 18 stone selections. The Theta torchiere stands 72 inches high.

Capri Lighting

The Smart Start electronic switching track connector converts new or existing Capri track systems to two-circuit control; it separates two groups of lights on the same track with only one two-wire supply connection. Smart Start increases control of the personal lighting environment by varying the amount and direction of light throughout the workday. Correct lighting for varied tasks can limit eye strain, fatigue, and glare, and help increase productivity.

Dazor Manufacturing

The Asymmetria desk lamp, manufactured by Oy Lival ab, Finland and available from Dazor, has a plastic head and adjustable, 34-inch steel arm. It uses and includes a 13-watt, 118-volt, compact fluorescent PL tube, and matches the light output of a 75-watt incandescent bulb, with 69.2 lumens per watt. Asymmetria is available in white or ebony.

Dazor Manufacturing

The *Asymmetria* desk lamp can be used for task lighting in CRT and other office areas. Its sleek, contemporary design complements virtually any office decor.

Eastrock Technology, Inc.

The *Reflectorized Optimum Performance Adapters* for compact fluorescent lamps are constructed of polished spun aluminum with a two-layer enamel finish. The combination of a photometrically designed reflector and the mirror-like finish, accelerate and drive the light far in excess of the lamp's standard lumen rating. They are available in five sizes and two lens arrangements. Optimum diameter allows for installation in any R30 or R40 recessed fixture.

The *Quad Adapter* allows for a 33 percent reduction in overall height compared to a normal compact fluorescent. The adapter operates all brands of 9-watt and 13-watt compact fluorescent quad lamps within manufacturers' specifications. It features a small diameter, high-efficiency and low-wattage consumption. Reflector covers are available for more downlight.

The Reflectorized Quad Adapter is designed to operate with 9-watt and 13-watt quad compact fluorescent lamps. The 9-watt adapted unit is 6 1/4 inches high and 9 inches in diameter, with an installed height of 5 1/2 inches. The 13-watt adapted unit is 7 1/4 inches high and 5 inches in diameter, with an installed height of 6 1/2 inches.

The Optimum Performance Adapters run cooler, use 75 percent less energy and increase bulb life more than ten times over the standard incandescent bulb. The compact design accepts all of the major manufacturers' compact fluorescent bulbs and allows the bulb to seat within the cas where the bulb-locking clamp assures a solid contact and the shortest possible profile.

The High-Pressure Sodium Adapter allows the retrofit of existing costly incandescent sockets with energy-efficient high-pressure sodium lamps simply by screwing the adapter into the socket. There is no need for rewiring or refixturing. It is also available in direct wire.

Exide Electronics, Lightguard Division

The Lightguard Series 2000 with Maxi-Power heads produce more than 2,000 candlepower each and consume only 12 watts of battery power. The high candlepower rating allows the unit to be mounted higher than units using PAR-36 lamps and still produce the footcandle levels required by NFPA-101 and OSHA. Greater spacing between units can reduce the installed cost of the project. The LEC-36 battery and Guardian charger have a 15-year maintenance-free service life, assured by the thick .310-inch positive plate, a low specific gravity electrolyte, and the accuracy of the charger.

General Electric Company

The compact halogen filament tube seen in the foreground is the heart of the Performance Plus TM halogen general service lamp. The 90-watt lamp replaces standard 100-watt incandescent bulbs and produces the same light output, but lasts twice as long. Rated lamp life is 2,000 hours compared to 750 hours for the standard 100-watt incandescent. It has flicker-free operation and the crisp white light characteristic of halogen. The outer bulb is of heavy glass construction and has a distinctive bottle shape.

GTE-Sylvania

The advanced features of the Designer 16 lamp include tungsten halogen capsule technology, a precision molded ceramic casing for beam control and aesthetics, and the incorporation of a diode in the lamp's electrical circuit to maximize efficiency and increase life.

The halogen capsule, combined with diode application, results in an energy efficient product with a long life of 2000 hours.

Herman Miller, Inc.

The High-Performance Task Light includes all the features of the Standard Task Light, plus a high-power factor, rapid-start ballast which draws less electricity than standard ballasts and operates at a cooler temperature. The fixture includes a specialized cool-white fluorescent lamp that offers better color rendition, longer life, and reduced electrical consumption up to 20 percent over standard fluorescent lamps. The fixture is designed for 60-Hertz, 120-volt, AC circuits. It is UL listed and CSA approved.

The Standard Undershelf Task Light includes a 6-inch deep batwing lens with translucent acrylic overlay that diffuses the light and directs it at optimum angles toward the work surface. Indirect glare is minimized. Light distribution is even and fixture brightness is reduced. The lens allows a high level of footcandles to be projected and increases output on vertical surfaces such as panels. Lightweight, spring-loaded brackets on the fixture hold it securely in place. The fixture is UL listed and CSA approved.

Holophane

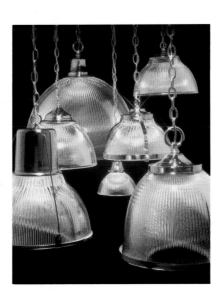

The Paradome R luminaire is a totally enclosed prismatic glass fixture that utilizes a low-wattage metal halide lamp to provide sparkle without glare, good color rendition, long lamp life and three times more light than incandescent. The luminaire is available in 100- and 175-watt lamp models, with standard stem length of 36 inches. The ballast tray mounts on T-bars above the ceiling. A variety of finishes are available.

Holophane

The Paradome luminaire is shown here with a white finish. For ease of maintenance, the glass enclosure, secured by torsion springs, can be removed by releasing the torsion springs.

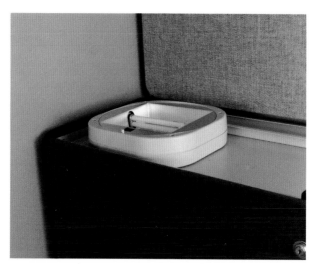

Shown here is the furniture-mount model of the Task-Mate portable task luminaire. The magnetic mounting system assures quick, easy installation. The user can put light where it is needed and control glare through the patented rotatable optical system. Colors available for all Task-Mate models include parchment, dark brown, black, grey, almond, warm brown 1, warm brown 2, fieldstone, light tone and innertone.

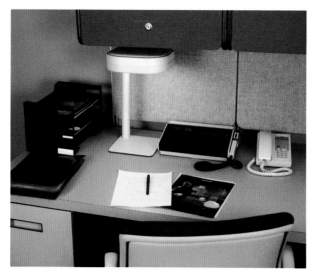

The Task-Mate TM, a user-controlled, portable task light system for the workspace, is available in desk stand and furniture/wall mount models. The task fixtures are made of injection molded plastic. The injection molded reflector is metallized for maximum specularity and reflectivity without the "rainbow" effect of metal reflectors. The fixture uses a 13-watt, 10,000-hour, miniature fluorescent lamp that does the work of a 20-, 30-, or 40-watt lamp because of the reflector and light control system. Shown here is the desk stand model.

Koch + Lowy Inc.

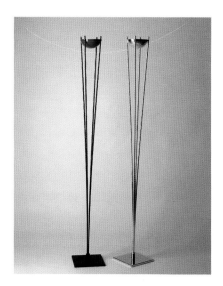

The Andrea (F-5180) luminaire, designed by Andrzej Duljas, is 74 inches high with a 7-inch shade and a 10-inch x 10-inch base. Maximum wattage is a 500-watt halogen bulb. It has a full range sliding floor dimmer. Available finishes include polished brass, chrome, black, or burgundy shade with black stems and base.

The Nuage (F-5010) luminaire, designed by Martine Bedin, is 84 inches high with a 36-inch shade and a 25-inch base. The luminaire comes with 2 x 250-watt double envelope halogen and a full range floor dimmer. Available finishes are frosted glass for the shade, and matte aluminum for the base and stem.

The Hawk Table (T-508), designed by Piotr Sierakowski, is 18 inches high with a 10-inch triangle base. The shade swivels 360 degrees, and the arm rotates 60 degrees and extends 26 inches. The finish is black Nextel. It uses a 50-watt halogen bulb. There is a high-low switch on the base.

The Scenist master controls combine four independent dimmers (channels) into one unit. These controls allow you to preset the dimmer intensities into four independent scenes or lighting situations. This lighting scheme is appropriate for a conference situation.

The four dimming channels consist of the following lighting groups: table downlights, accent spotlights, front wall wash, and side wall wash. Each channel may be independently controlled through the use of four raise/lower pushbuttons.

The light intensities are combined to create various effects that reflect the anticipated functions for the room. As these scenes are created, they are learned by the Scenist and are then re-created by pressing the appropriate scene select pushbutton.

The Lightolier Scenist Series are lighting controls engineered to combine functionality with esthetics. They are easily understood at a glance and utilize the latest computer technology.

Confusing control panels are no longer necessary. The Scenist Series contains self-explanatory controls with factory-supplied labels, that allow the user to identify essential functions in common language.

The transition between scenes is made at a 5 second fade rate. The transition may also be made instantaneously if desired, by pressing the desired scene button twice.

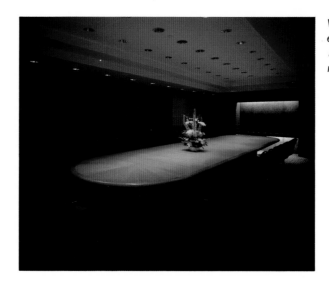

When the OFF push-button is selected, each channel fades at the 5 second rate, allowing occupants to exit while the room is still illuminated.

Lightscape Inc.

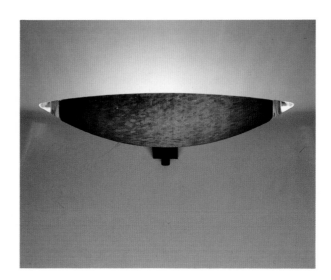

Flu, designed by Giusto Toso, is a wall bracket available in the color "marble polychrome." Flu is manufactured by Barovier & Toso.

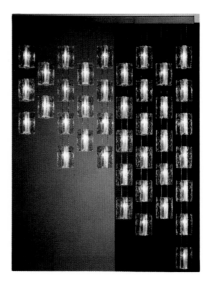

Giano Piccolo, designed by Luigi Chisetti, is a brushed strip crystal blind. Giano Piccolo is manufactured by Barovier & Toso.

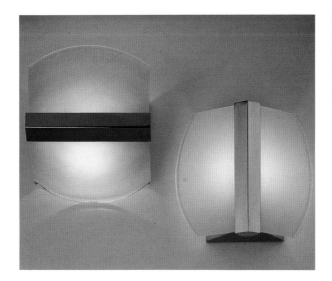

Sera, designed by Rudi Dordoni, is available in aquamarine and pink with external finishing in copper wall bracket. Sera is manufactured by Barovier & Toso, one of the leading glassworks in Murano, Italy.

Macro Electronics Corporation

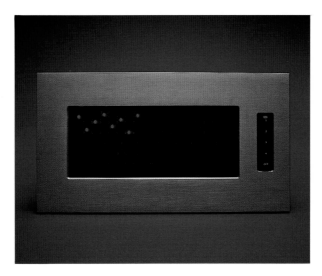

The Designer Preset System Master Station, suitable for use in board and conference rooms, is a four-scene, preset control panel with up to 15 channels. The station is easily located remote from the dimmer bank within the room that is being controlled, so the operator can see the lights when setting presets. The control station has a hinged door with hidden latch which covers the controls and leaves the preset select touch switches exposed for ease of operation. The door is available with a dark window for viewing the graphic intensity display. Timed preset transitions or crossfades are adjustable from 3 to 60 seconds.

Mark Lighting Fixture Co., Inc.

Note the lack of glare on the CRT screen with the use of the indirect ILS fixtures.

Mark Lighting
Fixture Co., Inc.

The ILS fixture is pendant mounted with spacers and butted on the wall. Note that the corner section is available with a downlight.

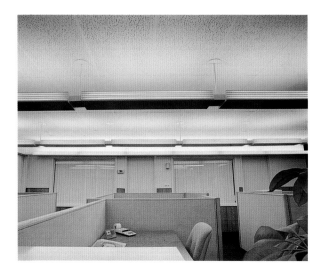

There is a soft glow on the edges of the ILS fixtures and even illumination on the ceiling.

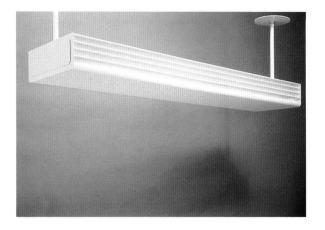

The ILS system offers matching solitaires for pendant, wall and task (portable) mounting. These allow the designer to maintain continuity within a space. Shown is the pendant-mount model.

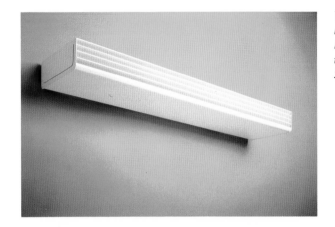

The ILS fixture contains an anodized reflector, and an open-design louver that broadens the pattern to make the transition from light to dark more even. Shown is the wall-mounted model.

North American Philips Lighting

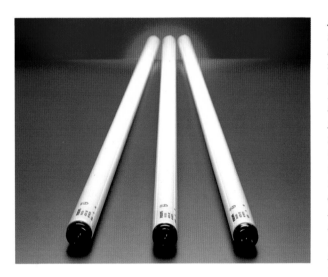

Advantage X TM line of 4-foot fluorescent lamps produces 3,700 initial lumens, yielding 17 percent more light than a standard F40 cool white fluorescent lamp. Its rated average life of 24,000 hours (at three hours per start) is 4,000 hours, or 20 percent, longer than standard F40 cool white lamps and means fewer replacements per year. The Color Rendering Index is 80. The T10 tube (1 5/6-inches) is easy to install and requires less storage space. Advantage X operates on all standard 40-watt rapid-start or preheat circuits. The line is available in three color temperatures which correspond to the appearance of standard warm white, white, and cool white fluorescent lamps.

Norton Industries

A dramatic, monolithic appearance is produced by the NWL-C-NM, a non-modular versionof the wood louver. Cell size varies from 3 inches to 24 inches on center. The standard is red oak, but a variety of hardwoods is available. The system can form a luminous ceiling, or an architectural ceiling utilizing downlight troffers and other fixtures.

The NWG-C Series, a non-modular wood grille ceiling design, is fabricated of solid hardwood. Both lighted and non-luminous designs create an appearance of continuity. Cell sizes are to 12 inches O.C. The fixtures are available in red oak, walnut, cherry, mahogany and other hardwoods.

The NWA wood arch designs accommodate ceilings up to 16 feet high and are available in module sized up to 4-feet x 4-feet. Though red oak is standard, a variety of hardwoods is available. All components are factory machined for easy on-site assembly.

The NFS wood fluorescent fixtures are available in module sizes from 1-foot x 4-feet, through 4-feet x 4-feet. Standard hardwoods include red oak, walnut, cherry and mahogany. Shielding media can be selected from a large variety of louvers, lenses and diffusers. Suspension methods include surface and pendant mount, grid or plaster ceiling.

New or existing 2-foot x 2-foot, 2-foot x 4-foot, and 4-foot x 4-foot fluorescent fixtures can be converted to a variety of wood fixtures through the use of the NRTC retrofit troffer converter. A variety of hardwoods, wood molding tapers and profiles are available.

The NLC coffer is a new addition to the company's family of solid hardwood coffers. The standard red oak coffer shown is 6 inches deep and tapered to simulate a skylight effect. The NLC deep tapered coffer is available in module sizes up to 4-feet x 4-feet, and a variety of hardwoods.

Norton Industries

Prescolite

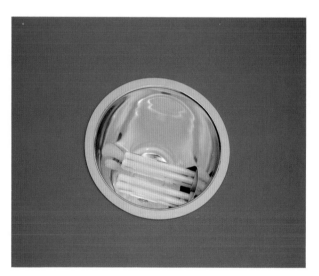

Compact fluorescent fixtures generally are used for applications involving low ceilings, or requiring only moderate footcandle levels. Prescolite's new high output compact fluorescent downlight, however, is suitable for high ceiling applications in banks, lobbies, stores and other areas. The dimming feature makes them attractive for use in meeting and conference rooms (such as the one above), and executive offices.

The new high lumen output compact fluorescent downlight fixtures accept either two 18-watt or two 26-watt quad tube lamps. Both versions feature a 7 7/8-inch aperture. Dimmable models are also available. Light output from the twin 26-watt downlight is approximately 3,600 lumens, which is comparable to a 200-watt incandescent fixture.

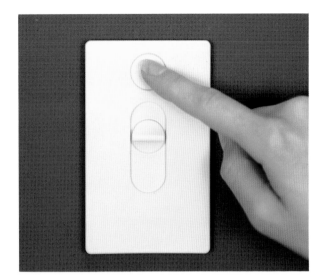

The new Horizon line of thin-profile wall-box slide dimmers can control loads of 600, 1,000, 1,500, or 2,000 watts. A top-of-the-line model provides full range dimming and power control of the same load from up to five different locations. A separate on-off switch above the slider allows the user to set lighting intensity and turn the lights on or off locally or from a remote dimmer.

Ron Rezek

Cygno is available in flat black with a translucent white or opaque grey shade. The base is 11 inches high. The arm spans 32 inches and raises 20 inches. The 13-watt fluorescent lamp (PL13) is included. The adjustable Cygno task light swivels, tilts, raises, pivots, and is non-glaring.

Zink Table is made with galvanized steel, polished copper, patina copper or satin black. It is 21 inches high, with a 14-inch diameter shade and a 3-inch diameter pole. The dimmer is included in the base. The 75-watt lamp (75 A/W) is not included.

Ron Rezek

Celeste is made with molded wire glass, a stainless ring, and brass rods. The luminaire uses four 13-watt fluorescent lamps (PL13)—included, or three 75-watt bulbs (75 A/W)—not included. Model 350 is available in two sizes: 23 inches in diameter x 40 inches high, and 23 inches in diameter x 18 inches high. Model 351 is also available in two sizes: 19 inches in diameter x 32 inches high, and 19 inches in diameter x 14 inches high.

Scientific Component Systems

The X18 series of compact fluorescent fixtures replace 75- to 100-watt incandescent, six-inch recessed downlights. Features include a patented zipcord for easy installation and a flush-to-the-ceiling trim ring that doubles as a cooling fin. When used to replace baffled 75-watt reflector floods, energy consumption drops by up to 76 percent, and illumination and uniformity increase. An electronic switching device that allows switching from one- to two-lamp operation can also be installed.

Steelcase Inc.

The Series 9000 system furniture integrates task and ambient lighting. Three fluorescent lamps are used in the indirect component and one lamp is used beneath the cabinets for task light.

The ECLIPSE task light allows the user to control the intensity of the illumination through an adjustable filtering mask that is rotated by hand.

Visa Lighting Corporation

The CB501B wall sconce has a white acrylic diffuser encircled by a four-bar trim. The trim is available in polished brass, natural brushed aluminum, or any desired painted finish. The standard model is 16 inches high and 7 inches wide with a 4 1/2-inch extension. A larger size model is also available that is 20 inches high and 9 inches wide with a 5 1/2-inch extension.

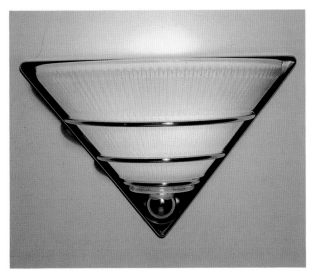

The CB2131 wall sconce provides uplight while transmitting light through clear acrylic cascades. Fixture body finishes include polished brass, polished chrome, or any desired painted finish. The fixture is 25 inches high and 16 inches wide with an 8-inch extension. Incandescent or fluorescent lamps can be used.

Visa Lighting Corporation

The CB931 is a one-piece, handspun, tiered wall sconce with a lensed bottom for downlighting. A closed bottom model is also available. Finishes include polished brass, polished chrome, or any desired painted finish. The fixture is 12 inches high and 18 inches wide with a 9-inch extension. The fixture can be made to accommodate incandescent, fluorescent or quartz lamps.

The CB2081 wall sconce provides wall-washing while transmitting illumination through bronze or clear acrylic stepped oval panels. Panels can be finished in polished brass or any painted finish desired. The fixture is 8 1/2 inches high and 18 inches wide with a 4 1/2 inch extension. It uses two 40-watt T-10 frosted lamps.

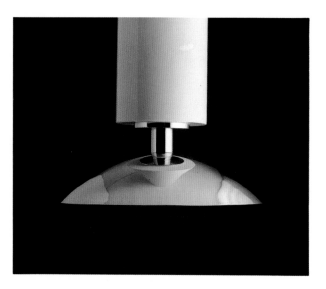

The DC121 pedestal-mount fixture provides uplight. Finished include polished solid brass, polished chrome or any desired painted finish. The fixture is 22 inches in diameter and 8 1/4 inches high from the base. The mounting base is 6 inches in diameter. Incandescent or metal halide lamps can be used.

The CB1300 wall sconce has a white prismatic glass half-cone complemented by polished, solid brass bars and backplate. Polished chrome or any desired painted finish is also available. The fixture is 8 1/2 inches high and 13 1/2 inches wide with a 6 1/4-inch extension. The fixture provides for incandescent or fluorescent lamping.

Waldmann Lighting

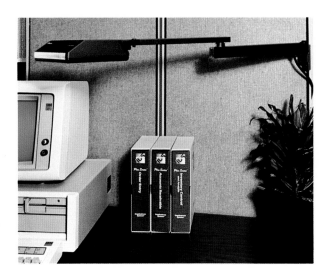

The 209 Designer Series task lighting is specifically designed for computer workstations, open office furniture systems, and CAD stations. The units allow the user to control the lighting. The built-in parabolic louver provides light control, and reduces glare and reflections from computer screens. The fixture arm can be moved vertically and horizontally, and the head can be swivelled. Two 9-watt lamps produce the same light as 80 watts of incandescent with 77 percent less energy and heat. Lamp life is approximately five years of normal office use.

INDEX

PHOTOGRAPHERS

INTERIOR DESIGNERS

ARCHITECTS

ELECTRICAL

DESIGNERS

LIGHTING CONSULTANTS

AWARDS

Albert S. Bard Awards for Excellence in Architecture and Urban Design

The Bard Award was established in 1963 by the City Club of New York, a leading civic organization. Although originally awarded to honor civic architecture, today any project built in New York City is eligible. Ada Louise Huxtable has called the Bard Awards "a barometer for the architecture of this nation."

The awards are named after the late Albert S. Bard, former trustee of the City Club of New York, who for 60 years fought vigorously for a better city. The jury, many of whom are outstanding professionals in the architectural community, is independent of the City Club. For more information, contact:

The City Club of New York
33 West 42 Street
New York, NY 10036

Edison Award

The Edison Award competition is open to lighting professionals who employ significant use of any General Electric lamps in a lighting design project. While the lighting installation need not be achieved exclusively through the use of General Electric lamps, the entry will be judged on the degree of achievement of certain criteria through significant use of GE lamp products.

The first prize is a customized Steuben crystal creation personalized with the winner's name. A distinctive plaque is presented to the owner of the installation. Similar plaques are awarded to entries reaching the final judging stage.

Entries will be judged on: functional excellence; architectural compatibility; effective use of state-of-the-art lighting products and techniques; good color, form and texture revelation; and energy and cost effectiveness. The panel of five judges represents the American Institute of Architects, the American Society of Interior Designers, the International Association of Lighting Designers, the Illuminating Engineering Society of North America, and the General Electric Company.

The competition was established in 1983. A descriptive brochure and entry form may be obtained by writing:

Edison Award Competition
General Electric Company
Nela Park # 4162
Cleveland, OH   44112

Halo/SPI National Lighting Competition

The competition is sponsored by Halo Lighting and held under the auspices of the American Society of Interior Designers. It was established in 1977 to honor designers who excel in the creative use of lighting as a basic element of design. Originality and technical ability are judged in designs using Halo Lighting Power-Trac track lighting, downlighting and SPI ambient/task indirect lighting.

Several awards are usually granted every year in each of two categories: First Place and Honorable Mention. Contact:

Halo Lighting
400 Busse Road
Elk Grove Village, IL 60007

International Illumination Design Awards Program

The Program, which is not a competition, provides an opportunity for public recognition of professionalism, ingenuity, and originality of lighting design based upon the individual merit of each entry judged against specific criteria. Criteria include considerations such as the complexity of the design problems and solutions and how well they deal with energy effectiveness, environmental limitations, costs, originality, and suitability for and consistency with aesthetic and functional requirements.

Judges are selected from a broad professional spectrum which represents knowledge of lighting and design excellence. The awards are presented on several levels, beginning with local and regional. The highest honors include the Edwin F. Guth Memorial Special Citation, Award of Excellence and Award of Distinction. The Excellence and Distinction awards consist of one glass sculpture per design team, and one plaque per owner. All other awards are certificates.
For further information, contact:

Illuminating Engineering Society of North America
345 East 47th Street
New York, NY   10017

Lighting Design Awards Program

The Awards Program was begun in 1983 to increase awareness of good lighting design. Projects must demonstrate aesthetic achievement, technical merit, and sensitivity to the architectural concept. Certificates are presented for Awards of Excellence, and Honorable Mentions. Nonmembers as well as members of the International Association of Lighting Designers may enter. For more information, write:

International Association of Lighting Designers
18 East 16th Street, Suite 208
New York, NY 10003

DATE DUE